ENJOYING MOTHS

ENJOYING MOTHS

Roy Leverton

With line illustrations by
Michael Roberts

T & A D
POYSER
NATURAL
HISTORY

This book is printed on acid-free paper

Text © Academic Press, 2001
Photographs © Roy Leverton
Line art © Michael Roberts

First published in 2001
by T & A D Poyser Ltd
32 Jamestown Road, London NW1 7BY
http://www.academicpress.com

ISBN 0-85661-124-7

Library of Congress Catalog Number: 00-107416

A catalogue record for this book is available from
the British Library

Designed and typeset by Peter Champion
Printed and bound by Grafos SA Arte Solore Papel,
Barcelona, Spain

01 02 03 04 05 06 GF 9 8 7 6 5 4 3 2 1

Contents

Preface

This book is aimed at those who might wish to learn more about moths. Perhaps they consider themselves to be good general naturalists; maybe they already specialise in a different field. That would be the normal pattern. Few people consciously set out to make moths their main hobby: an interest in this group more often develops indirectly, through casual encounters while pursuing other branches of natural history.

For the ornithologist, the caterpillars of moths come to notice as a major prey of insectivorous birds, so important for nestlings that tits synchronise their breeding season to coincide with this flush of food in spring. Adult moths form part of the diet of exciting birds such as the Nightjar and Hobby. The twitcher, ever hoping for rare vagrants, knows that moths migrate too, using the same weather systems. Accordingly, most bird observatories also run a moth trap.

For those interested in mammals, the link between bats and moths is obvious, but hedgehogs, shrews, mice and voles also eat them, especially during the pupal stage. So do humans in some parts of the world, but fortunately not in Britain. Australian Aborigines harvest aestivating Bogong Moths *Agrotis infusa*. Native American tribes considered White-lined Sphinx *Hyles lineata* caterpillars a delicacy, and used them as the basis for ceremonial stews (Watson and Whalley, 1975).

For the botanist and the gardener, moth caterpillars are the most obvious cause of leaf damage. Plants have evolved many defences, both chemical and physical, against these depredations, while at the same time encouraging moths to act as pollinators. In response, moths have found cunning ways to overcome these defences, or turn them to their own advantage.

However, the most common route towards an interest in moths is via butterflies, their more showy daytime cousins. As Young (1997) rightly stresses, a butterfly is just a moth by another name. And Britain has relatively few butterflies: with moderate luck and a bit of effort it might take only a couple of years to see all the species found in a county, and not much longer to 'tick' every resident butterfly in Britain. What then? Moths, in contrast, offer endless scope. With roughly 30 species for each butterfly, who could expect to see them all? Even after a lifetime's study, it is realistic to hope for new moths every year, perhaps without travelling further than one's own garden.

Nor should those who live in urban areas despair – moths are found everywhere. My own childhood was spent in Salford, part of the Manchester conurbation. Air pollution from heavy industry was rife, and the countryside far out of reach. There was so little wildlife that I envied my schoolmates whose terraced houses had rats, and asked whether I might be permitted to come and study them. This was the Salford of Lowry. Once, the artist himself treated our school to a well-rehearsed version of his life story while trusted pupils fed him prearranged questions. What we would really have liked to ask him was why he painted Salford anyway. Art to us meant landscapes, galleons, nymphs surprised while bathing, and great cities like Paris, Rome and Venice, not industrialised slums.

Yet even in that desperate environment there were moths, with the added interest that so many had evolved dark (melanic) forms to match their soot-polluted background. Thus began a lifelong passion that has never faded. Why should it, when moths are so fascinating? And long-awaited species can turn up almost anywhere, when least expected. Perhaps the most memorable of my own first sightings was the Privet Hawk *Sphinx ligustri* seen from a moving bus, bringing my journey to an unscheduled halt. What a magnificent moth! Unfortunately, it was only my third day in a new job – not a good time to arrive half an hour late. In those days there was little protection: firms could, and did, sack employees on whim. The foreman listened impassively to my stammered explanation; it sounded a bit lame even to me. The silence that followed seemed unending. Finally he delivered his verdict. 'You got off the bus because you saw a *moth*? That's the daftest excuse I've ever heard – it must be genuine! Make up the lost time in your lunch hour.'

Moths generate many such memories. As a subject for study they are inexhaustible. They lead the keen observer from the seashore to the tops of the highest mountains, through marsh and forest, across heath and moor: all environments have their characteristic species. And so does suburbia. Because moths are active at night, they make a good hobby for those forced to spend the day at school or work. What better way to unwind after a hectic spell in the office than wandering round the garden at dusk, drinking in the scent of the flowers, and watching moths as they wake to fly and feed? What other branches of natural history can be pursued even while one sleeps – by running a light trap?

My own interest might not have taken root without the encouragement of my parents and the generosity of my primary school headmistress, a Miss Grimshaw, who gave me South's (1907–09) identification guide from her own library. I used those two volumes so much they fell to pieces. WHT Tams, AL Goodson, Allan Brindle, Bernard Kettlewell and EB Ford's secretary patiently answered my childish letters. In Sussex, the late TG Rea and family allowed me to wander anywhere over their land by day or night for 25 years. Colin Pratt stimulated my interest in accurate recording, and Dennis Dey introduced me to the effectiveness of modern light traps. Since I moved to Banffshire, Mark Young, David Barbour and Bob Palmer have done their best to show me all the choicest Scottish moths, and uncomplainingly answer endless queries. Correspondents too numerous to name have willingly provided me with information over the years. I particularly thank Gerry Haggett for teaching me to appreciate the early stages and the fascinating behaviour of caterpillars, and for helping me to formulate my ideas.

Like every lepidopterist, I owe a special debt of gratitude to my family for their tolerance. Laurie, my wife, taught me photography, and now puts up with moths or caterpillars in almost every room of the house. My elder son David sorts out all the problems caused by my incompetence with computers.

In this book I have tried to convey some of the enjoyment to be gained both from moths and from mothing. Mark Young, as always, has cast his professional eye over parts of the draft, and saved me from numerous errors. Otherwise, with no career or reputation to worry about, I have shamelessly taken advantage of my amateur status, and written about topics that interest me in the hope that they will interest others.

Finally, without the skill and dedication of Mr Koruth and the medical and nursing staff of Aberdeen Royal Infirmary, who looked after me for many months as if I was some particularly rare and delicate caterpillar, this book might not have been written at all.

Chapter 1

What is a moth?

Moths and butterflies belong to the Order Lepidoptera, characterised by the covering of scales (modified hairs) that overlap like roofing tiles on both surfaces of their relatively large wings. Because of their attractive appearance the Lepidoptera are very popular with collectors, and this has helped them to become perhaps the best known of all the insect groups. Estimates vary as to the worldwide number of species. Young (1997) quotes a figure of 130,000 known Lepidoptera, of which about 12 per cent are butterflies, whereas Duckworth (1975) gives the number of described species as approximately 165,000, about 10 per cent of them being butterflies.

Ninety-six per cent of British Lepidoptera are moths

Currently, the list of Lepidoptera ever recorded in Britain stands at 2,600 species (Bradley, 1998), give or take a few. For various reasons, it is impossible to be more precise. Some of the older records are dubious, or probably misidentifications; occasionally species are lumped or split as more is learnt about them; new moths are added annually either as migrants, colonists or overlooked residents. Of the 2,600 species, 96 per cent are moths. Even our 106 butterflies include about 24 adventives (species accidentally imported by man with vegetable produce or other goods), a further 12 species recorded so infrequently that they are clearly only vagrants or strays, and two or three that have been extinct here for many years. This leaves fewer than 70 butterflies that can be considered either resident in Britain, or regular if scarce migrants. Of course, these same caveats apply to the 2,496 moths on the list as well, but to a lesser extent, because stray exotic butterflies are relatively more likely to be reported by the general public. Emmet (1992) considered that about 130 (just over 5 per cent) of the moths on the British list were adventives.

Because only 4 per cent of Britain's Lepidoptera are butterflies compared with global estimates of 10–12 per cent, we might be judged particularly well-endowed with moths – or else poor for butterflies. Since Britain is a small island not renowned for the sunny weather that butterflies need, there may be a grain of truth in the latter view. However, even in Europe as a whole, butterflies only comprise about 5 per cent of the 8,470 known species of Lepidoptera (Karsholt and Razowski, 1996). Their apparently higher proportion globally may be an artefact. Over vast areas of the tropics and in other sparsely inhabited regions, butterflies have been by far the better studied. Perhaps 90 per cent of the world's species are known (TB Larsen, personal communication), whereas we may surmise that less than half of the world's moths have been described and named, particularly the smaller ones.

So what is the difference between a butterfly and a moth?

This is one of the most frequent and tiresome questions that the enthusiast has to face, especially as the correct answer is 'not very much'. We give the name moths to the more primitive (in the biological sense of 'arising first') families into which the Lepidoptera are divided – and also to the later, supposedly more 'advanced' ones. Sandwiched between them is a group of families which collectively we call butterflies. Admittedly, these families are rather distinctive ones that share many features in common: their members have clubbed antennae, they tend to be brightly coloured, and they are almost entirely diurnal. They also lack the frenulum, a catch and bristle arrangement linking forewing and hindwing (Figure 7) found in most moths. But basically it is merely a question of semantics, equivalent to asking the difference between a shark and a fish, or a hawk and a bird.

Micros and macros

More controversial is the division of the Order into the microlepidoptera and the macrolepidoptera. This has very little scientific validity, and it has been customary for well over a hundred years to deplore the practice. It has survived in spite of this because of the overwhelming merit of practicality. Basically, the butterflies and all the families of (generally large) moths that come after them in the classification scheme are known as the macrolepidoptera (macros for short). All the families of moths that come before the

Plate 1: The Gold Swift *Hepialus hecta* is a primitive moth, but considered an 'honorary macro' because of its size. The male's metallic markings, conspicuous while hovering, may have a sexual function. (Banffshire).

butterflies are known collectively as the microlepidoptera or micros, because the great majority of the moths in them are indeed very small. This neat division is somewhat blurred in that several of the more showy micro groups such as the swift moths (Hepialidae) (Plate 1), the burnets (Zygaenidae) and the clearwings (Sesiidae) are usually counted as honorary macros. Partly this is done for historical reasons, as burnets and clearwings were once thought to be allied to the hawk-moths (Sphingidae). Note that the distinction is drawn at the family level, rather than on the size of the individual moth. This means that a few of the macros are in fact considerably smaller than many of the moths that are classed as micros.

Clearly the division is an artificial one, simply a question of convenience. The micros are just as much moths, and equally fascinating in their own way, as the macros. Yet they remain relatively neglected, the province of specialists. There are many parallels in other groups. Few arachnologists tackle the innumerable small black money spiders in the family Linyphiidae. Many knowledgeable amateur botanists do not even attempt to come to grips with the grasses, rushes and sedges. And what could be more artificial than the botanist's categorisation of some species as trees, while others even in the same genus are classed as shrubs or herbs?

Scientific names

The prerequisite of all studies of living organisms is to be able to identify by name the species in question. Whereas all the British macrolepidoptera have English names in general use, very few of the micros do, apart from some of the more obvious species and those that are well-known pests. For the rest, there is no alternative but to refer to the moth by its scientific name. Inevitably, for many beginners this acts as a deterrent. In general, scientific names are not user friendly, especially as few people today possess enough knowledge of classical Latin and Greek to understand what they mean and how to pronounce them. The situation is not helped by the attitude of superiority adopted by a few experts towards those who lack their esoteric knowledge – though this is rare indeed among those who study moths. However, I once encountered it from a botanist who persisted in referring to every plant by its scientific name, causing me the embarrassment of continually having to ask him to translate. In the end, I set about learning the scientific names of all the relevant plants on the reserve, only for the botanist to revert to using common names once his game was no longer fun.

Curiously, there is a difference between the way botanists and entomologists use scientific names in conversation. The former tend to use only the generic name, the latter only the specific. Thus to point out a Fox Moth *Macrothylacia rubi* caterpillar feeding on a bramble *Rubus fruticosus*, it would be customary to say 'Look at that *rubi* on the *Rubus*', never 'that *Macrothylacia* on the *fruticosus*'. A lepidopterist would only use the generic name on its own if confident of the genus but uncertain as to species of the moth in question. Interestingly, birders likewise rarely use scientific names in everyday speech unless unsure of the identification, as when they refer to 'an odd-looking *Sylvia* warbler' or 'a *Phyllosc(opus)* with a wing-bar'.

For the larger moths, there is a choice: either their English (vernacular) name can be used, or their scientific one (which is made up of a generic name usually taken from a Greek root followed by a Latinised specific name). Each side has its supporters. It can be argued that scientific names are universal, and enable lepidopterists anywhere in the world to understand which species is being discussed, whereas English names apply in Britain only (for example, our Ni Moth *Trichoplusia ni* is called the Cabbage Looper in North America). Unfortunately, because the rule of precedence applies for scientific names, workers in different countries do not always agree which is the earliest valid one. The moth known in Britain as The Uncertain *Hoplodrina alsines* is called *H. octogeneria* elsewhere in Europe, and there are various similar cases. In fact, scientific names are surprisingly inconstant, and those using them must be prepared sometimes to abandon one that has been in use for decades or even centuries, and learn to call the moth something else. As many of our macrolepidoptera were first described in the latter half of the eighteenth century, it would be reasonable to expect that their correct specific names would have been sorted out by now. Unfortunately, there are a few sad people whose hobby it is to overturn long-familiar names by uncovering obscure earlier ones, or by casting doubt on whether an old description does indeed refer to the species in question when the original type specimen is no longer in existence. Nearly a century ago Tutt (1902) criticised those 'who study scientific names rather than the insects themselves'.

Whilst it is possible to hope that specific names will one day become fixed, the same cannot be said for generic names. These depend, not solely upon precedence, but upon our judgements of how closely species are related to each other. Whereas the species is largely a biological concept, the genus is much more a human invention, an attempt to categorise, to pigeonhole, and to create some sort of order and system, based on our view of the degree of kinship between different species. As such, it is inevitable that this view will often change as research increases our understanding, or new methods of investigation, such as DNA analysis, are brought to bear. But it should not be forgotten that the function of a genus is as much to group as it is to divide, and the increasing tendency to dismember familiar genera because of rather minor differences among the species they contain is regrettable.

English or vernacular names

The law of precedence does not apply to vernacular names, and mercifully so – if it did, we would still be calling the Small Copper *Lycaena phlaeas* 'the small golden black-spotted Meadow Butterfly', as it was named by Petiver in 1699. The English names of British moths were effectively stabilised when Richard South published his illustrated guide in 1907–09. Since then, there have been only a handful of minor changes, usually to avoid ambiguity. South gave the name Muslin to two species, *Diaphora mendica* and *Nudaria mundana*, so the latter is now known as the Muslin Footman to distinguish it. Likewise, two moths were called The Flame in his book, one of which is now the Ruddy Carpet *Catarhoe rubidata*. Perhaps South, in the politer society of Edwardian England, was reluctant to use such an adjective! The Satin Carpet *Tetheella fluctuosa*, not being a geometrid like all the others known as carpet moths, is now the Satin Lutestring, while the Lesser Satin has become the Common Lutestring *Ochropacha duplaris*. Otherwise, virtually all of South's vernacular names have survived untouched, whereas the scientific names that were then in use have been amended greatly, sometimes several times.

A few of the stuffier lepidopterists profess to find the English names of moths too fanciful to be taken seriously. Others they consider to be simply inappropriate. There is the feeling that scientific names must by their very nature be more accurate, less whimsical and, well, more *scientific* than the vernacular ones. It has become much harder to embrace this view since the publication of a fascinating guide to the history and meaning of the scientific names of British Lepidoptera (Emmet, 1991). This has revealed that many generic and specific names are quite as far fetched, poetical, inappropriate or banal as the English ones, if not more so. Sometimes they perpetuate the same error: the Golden-rod Brindle *Lithomoia solidaginis* has nothing to do with golden rod *Solidago*, nor the Horse Chestnut moth *Pachycnemia hippocastanaria* with the tree *Aesculus hippocastanum* whose name it bears. Of course, there are instances where the scientific name does

Plate 2: Small Elephant Hawk-moth *Deilephila porcellus*. Piglet, pachyderm or bird of prey – as so often, both scientific and English names are equally fanciful. (Sussex).

capture the character of the moth much more accurately than the English one, but this is by no means always the case. I particularly like the description of the Small Elephant Hawk-moth *Deilephila porcellus* as 'the evening-loving little pig' (what could be more fanciful than that?). This moth (Plate 2) always reminds me of a plump pink piglet as it suckles at campion flowers at dusk – although Emmet states that the name actually refers to the caterpillar.

So we should not regard scientific names with too much awe. Nor should we consider the vernacular names inferior. Admittedly, many are quaint and archaic, dating back to Georgian times, coined perhaps by richly brocaded gentlemen of leisure in an age when liveried servants carried the collecting apparatus and rustics tugged their forelocks as the cavalcade went past. Like Marren (1998), I believe we should cherish them as part of our heritage. Names like Green-brindled Crescent, Merveille du Jour and Cream-spot Tiger wove a spell over me as a child that has never been broken, as they must have done to countless others. Vernacular names draw people in, whereas scientific names seem almost to have been designed to discomfort and exclude beginners.

Nevertheless, in the end it is best to be familiar with both. Scientific names show the generic affinities between species when the English names do not, as in the case of the Blossom Underwing *Orthosia miniosa*, Common Quaker *O. cerasi* and Clouded Drab *O. incerta*. English names are better for conveying the general appearance of moths which are not closely related enough to belong to the same genus, as with those ending in Carpet (22 different genera in the subfamily Larentiinae alone), Hawk-moth (15 genera) or Wainscot (9 genera). Sometimes scientific names, far from being long and unwieldy, are easier to use: it is certainly quicker to shout *melanopa* than 'Broad-bordered White Underwing' – by which time that fast-flying diurnal species would long be out of sight. But perhaps the greatest advantage of learning both English and scientific names is that it doubles the chance of remembering the name for those of us whose memory is no longer quite as good as it was!

Similarly, there is no need to be put off by technical terms and scientific jargon. Often, the English word is not only easier to understand, but more accurate as well. Many lepidopterists use the term 'ovum' (plural ova) instead of egg. In fact, an ovum is strictly a female germ cell. When this is surrounded by food for the developing embryo and encased in an outer shell, the whole package becomes an egg. Moths, like birds, lay eggs not ova! Likewise with the term 'larva': many other insects have larvae, as do a wide range of marine invertebrates. 'Caterpillar', on the other hand, refers strictly to the larva of a butterfly or moth, and thus is far more precise. However, as the term 'chrysalis' only properly applies to butterflies, for moths there is no English equivalent of 'pupa', the Latin for doll. (The arms and legs of early dolls were fused with the body, in the same way that the future moth's appendages are.) It may also be better to refer to the 'imago' (plural 'imagines') than the 'adult moth' or 'perfect insect' – especially if the latter is a bit worn and tattered.

The life cycle of moths

As every schoolchild used to be taught (and perhaps still is) the life cycle of the Lepidoptera is a classic example of complete metamorphosis. The female mates and then lays eggs, which hatch into caterpillars. These caterpillars eat the leaves of plants, and grow until their skin is too tight to stretch further, whereupon the skin splits to reveal a new and looser skin beneath. After several changes of skin, the caterpillar reaches full size and turns into a chrysalis or pupa, from which the adult insect emerges in due course. This basic pattern holds good for the vast majority of moths, in spite of their diversity. There are relatively few exceptions, and these are mainly among the micros.

Why lay eggs?

Virtually all moths do lay eggs, but this stage is missed out in a few *Coleophora* species such as *C. albella*, where the female deposits fully formed caterpillars directly onto the food plant, Ragged-Robin. This ability to give birth to live young is so widespread in the animal kingdom that the great mystery is why no birds are viviparous, like some of the reptiles from which they evolved. Why should penguins stand on their ice floes for months, with an egg cradled on their feet, when it could be safely retained within the body? For short-lived moths, the drawback could be the risk of predation during the gestation period, whereas eggs can be laid immediately after mating, and often they are.

For nearly all moths, the eggs need to be fertilised by a male before they are laid. Parthenogenesis (the ability of females to reproduce asexually) is common in some insect groups, like the aphids and the bees, wasps and ants, but rare in moths. In Britain, some members of the Nepticulidae are known to be parthenogenetic, such as *Bohemannia pulverosella*, *Ectoedemia argyropeza* and *Stigmella microtheriella*, and it is a feature of the Psychidae. In this strange group the males are fully winged and active, but the females are little more than bags of eggs, and never emerge from their cocoon. Some, like those of *Epichnopterix plumella*, lack not only wings, but legs and antennae too; in *Acanthopsyche atra* the female does not even have eyes. There is an account in Heath and Emmet (1985) of a brilliantly intuitive experiment whereby females of the latter species were fed to a captive Robin. Two weeks later, dozens of healthy newly hatched caterpillars crawled from the Robin's droppings, the eggs having survived their passage through the bird's gut in much the same way as the seeds in berries do.

Size versus quantity

The number of eggs that a moth can lay varies between species. Generally, the size of the egg is roughly proportionate to the size of the moth, but there is a choice between producing fewer but bigger eggs, or larger numbers of smaller ones. Species that go for size rather than quantity include the Kentish Glory *Endromis versicolora*, where females lay not many more than 200 eggs in spite of having bulky abdomens. Disproportionately large eggs are also produced by two inhabitants of bleak mountain summits, the Northern Dart *Xestia alpicola* and the Broad-bordered White Underwing *Anarta melanopa*. Crowberry is the food plant of both these species, so perhaps their caterpillars need to be large enough on hatching to tackle the tough glossy cuticle of its leaves. Moths which produce great numbers of relatively small eggs for their size include the Large Yellow Underwing *Noctua pronuba* and the Sword-grass *Xylena exsoleta*, with 1,500 or more. Clearly, mortality rates in the early stages must vary considerably between species adopting these different strategies.

Laying the eggs (ovipositing)

Moths do not have too much scope for diversity in the matter of egg laying. Unlike birds, which can transport food to helpless nestlings or lead chicks to good foraging areas, moths must lay their eggs where the newly hatched caterpillars will be able to find their own food quickly. Thus eggs are normally laid on the food plant itself, or close by. Hiding the eggs from predators, or protecting them with a covering of irritant hairs (Plate 3), seems to be the limit of parental care in moths. No moth is known to guard its eggs and offspring, as do a few other invertebrates such as spiders, bees and ants. Indeed, virtually all moths will be dead before their eggs hatch.

The way the eggs are laid also varies. Those of the swift moths are simply dropped while the female is in flight, presumably over suitable habitat, and the Oak Eggar *Lasiocampa quercus* also uses this technique. The Chimney Sweeper *Odezia atrata* likewise broadcasts its unusual, brown, seed-like eggs, which overwinter amongst the ground litter and hatch 8 months later to greet the new spring growth. This explains how a colony in a Sussex orchard was able to thrive even though the Pignut on which the caterpillars fed was mown and raked up at the end of each summer.

The great majority of moths deposit their eggs on the caterpillar's food plant. Often, the particular scent or feel of this plant is needed to stimulate the female to lay. Most moths are fairly precise in choosing where to lay their eggs – whether on the leaves, stem, buds or flowers. Species with a long ovipositor can push this into small cracks in bark, or into other slits and crevices, and so hide the eggs from view. Given the right conditions they can do this so successfully, even in captivity, that the inexperienced observer may easily assume that the female has failed to lay at all. I have had to demolish dry grass stems strand by strand to reveal the rows of eggs laid by a female Crinan Ear *Amphipoea crinanensis*; the eggs are covered with a sticky substance which cements shut the crease or fold in which they are laid. As the egg stage lasts 7 months or more in this species, such a safety measure is no doubt worthwhile.

Plate 3: Dark Tussock *Dicallomera fascelina* female with an egg batch – densely covered by dark grey protective hairs from her tail tuft. (Banffshire).

Batch size

Moths also show a choice of strategies regarding how many eggs to lay in one place. The eggs may be laid singly, or in batches of various sizes. Each species is fairly consistent in this respect. The Common Lutestring places a single egg on the tips of the serrated teeth around the edge of a birch leaf, where they are easily overlooked. Most of the prominents (Notodontidae) lay only one or two eggs before moving on. Obviously, moths with caterpillars that feed communally, at least while they are small, must lay their eggs in groups of the appropriate size. Thus the Kentish Glory lays batches with an average size of about 18 eggs (Plate 87), and the caterpillars remain together until they are about half-grown (Plate 4). Emperor Moth *Saturnia pavonia* caterpillars are also communal at first, and those of the Buff-tip *Phalera bucephala* stay together until nearing full growth (Plate 115).

However, many species whose caterpillars are not communal also lay their eggs in large batches, but here the caterpillars are endowed with the instinct to disperse as soon as they have hatched. In captivity at least, those of the Sword-grass crawl rapidly for up to 3 days, before suddenly losing this urge and becoming sedentary. Other species depend upon the wind for dispersal. Caterpillars of some tussock moths (Lymantriidae), such as the Vapourer *Orgyia antiqua* and the Gypsy Moth *Lymantria dispar*, have long hairs that catch the breeze, and which might even carry the latter species over the English Channel (Bradley, 1998). Those of the Peppered Moth *Biston betularia* (and probably various other species in the same subfamily, Ennominae) copy the technique used by many young spiders, of letting out a silk thread until it is long enough to act as a parachute allowing them to become part of the aerial plankton (Kettlewell, 1973). No doubt it is adaptive that all the species mentioned can feed on an exceptionally wide range of trees and shrubs, as presumably they have no control over where they land.

Clearly, there are costs and benefits whatever strategy is adopted. Laying eggs singly, and choosing each site carefully, spreads the risk and gives the newly hatched caterpillars their optimum chance of survival. However, it costs the female

Plate 4: Kentish Glory *Endromis versicolora* caterpillars are gregarious when young, clustering at the end of twigs in this characteristic manner, perhaps to deter predators. (Moray).

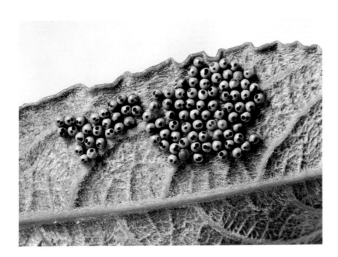

Plate 5: Buff-tip
Phalera bucephala
eggs, parasitised by a
Telenomus wasp so
tiny that one can
develop within a
single egg. Almost
every shell shows the
neat round hole
chewed by an
emerging adult.
(Sussex).

energy and time, so that she might succumb to exhaustion or be predated before all her eggs are laid. If too choosy, she might not find anywhere that stimulates her to lay at all, and die still gravid; this certainly happens with captive females in spite of the best efforts of breeders. Laying the eggs in large batches, on the other hand, saves the female time, but increases the risk that all her output will be lost in one fell swoop (Plate 5). Perhaps unsurprisingly, compromise is common: most moths lay their eggs in small or medium-sized batches, and are only moderately particular where they lay. Often, they give themselves a chance of leaving at least some progeny by laying a batch of eggs immediately on separation after mating. In the wild, I have observed this in species as diverse as Lunar Hornet Moth *Sesia bembeciformis*, Emperor Moth and Dark Tussock *Dicallomera fascelina*, and Tutt (1901–05) mentions it as a common habit of the tiger moths (Arctiinae).

Hatching

Inside the egg, the caterpillar may begin to develop immediately, or it may only develop after a dormant phase, as with many eggs that overwinter. The caterpillar emerges by eating a hole in the egg's chitinous shell. Depending on the species, the hole need only be large enough to allow the caterpillar to escape, or the eggshell may be eaten as its first meal. After this, the caterpillar cannot survive long unless it can find suitable food quickly. Most have very limited reserves, and die within hours of hatching from exhaustion and dehydration if the right food is not available.

Food for the caterpillar

The vast majority of species feed on one or more of the higher plants. The Lepidoptera are a relatively young order in evolutionary terms. The first undisputed fossil record – a few wing scales preserved in amber – dates back about 100,000,000 years, whereas primitive land plants were flourishing 350,000,000 years ago. It was only when flowering plants began their rapid evolutionary advance in the middle of the Cretaceous period (about 80,000,000 years ago) that Lepidoptera too began to increase and diversify to take advantage of this potential food supply (Watson and Whalley, 1975).

Inter-relationships between moths and plants

Because plants and moths evolved hand in hand, there is something of a love–hate relationship between them. Like bees, moths are bribed by nectar to visit flowers, and inadvertently cross pollinate them. In a few extreme cases, plant and moth depend entirely upon each other. The best known example is the Yucca plant of North America, pollinated only by micro-moths of the genus *Tegeticula*, whose caterpillars feed on the developing seeds. In Britain, the caterpillars of several moths in the genus *Hadena* feed on the seeds of campions that the adults have helped to pollinate. Then there is the remarkable story of an orchid endemic to Madagascar with a nectary so long that Charles Darwin reasoned that there must be a hawk-moth not known to science, with a proboscis of at least 12 inches (30 cm), able to reach the nectar. Sure enough, Darwin's predicted hawk-moth was discovered in due course, and appropriately named *Xanthopan praedicta*! Rather spoiling the story, it was later found to be only a subspecies of *X. morgani* from the African mainland.

However, the relationship between moth and plant is not always mutually beneficial. There are many moths, including whole families, where the adult lacks a functioning proboscis and does not feed. Hence it cannot help to pollinate flowers. Nor do all plants require insects to act as vectors: many trees and grasses are pollinated by the wind. In both these cases, it is hard to see how the plant could derive any benefit from the moths whose caterpillars feed on it. Consequently, even plants not normally regarded as being poisonous have evolved a wide range of chemical defences against being eaten. These include alkaloids, glycosides, oxalates and tannins (Cooper and Johnson, 1984). Often, any leaf damage stimulates the plant to increase production of its toxins. Moths have responded in various ways. Caterpillars may either bypass the plant's defences, evolve tolerance to the toxins, metabolise them into harmless compounds, or even sequester them for their own use. This subject has been well researched because of its commercial importance – some caterpillars are pests of farming, forestry or horticulture – but it is outside the scope of this book. Young (1997) gives an excellent summary.

Although we have to assume that plants' chemical defences do actually work against caterpillars, either by minimising damage or reducing the number of species that can feed on a particular plant, clear-cut instances are hard to find. Both in the wild and in the laboratory, caterpillars reject plants which are not part of their normal diet, and may even prefer to starve rather than eat them, but this does not necessarily mean that the plants would prove toxic if they did. The Yew is considered by some to be the most poisonous plant in Britain because all parts except the fleshy fruit contain a complex mixture of alkaloids, plus a cyanogenic glycoside and an intensely irritating volatile oil. Certainly it has almost no moths associated with it in

Britain, apart from the Satin Beauty *Deileptenia ribeata*. Ash and privet, again full of glycosides, support few caterpillars, and those are mainly specialist ones, as would be expected. On the other hand, oak is second only to sallow in the number of species that feed on it, in spite of its high tannin content. Knot-grass, famed among moth breeders as an almost universal food plant for caterpillars that feed on low-growing plants, none the less contains oxalates.

It is not surprising when moths that have evolved to feed on certain specific plants are able to cope with those plants' defences. More remarkable is the ability of the polyphagous species (those that can use a wide range of food plants) to overcome the correspondingly broader spectrum of chemical defences, and to perform equally well on food that presumably has different nutritional values as well as physical characteristics. Starting with tiny caterpillars from the same batch of Sword-grass eggs, I reared 30 entirely on dock, and 30 entirely on Bird Cherry, side by side under identical conditions. The aim was to determine which was the better food plant for this species. The result was totally unexpected. There was not the slightest difference between the two batches at any stage. The caterpillars grew at exactly the same rate, reached exactly the same size, and their survival rate was identical too, with 26 live pupae from each batch. Dock and Bird Cherry are very dissimilar, the former having fleshy leaves with a high water content, the latter much tougher ones that contain prunasin, which yields hydrocyanic acid when the leaves are crushed. Apparently, the caterpillars were able to adjust their intake or their metabolism to grow at the same rate in spite of differences in the food plant. If this is the case, we should regard polyphagous species as more, rather than less, sophisticated than the specialists that are tied to a particular plant. A car engine that could perform equally well on paraffin, diesel or petrol would be far harder to build than one designed to run on a single fuel.

Feeding strategies

Like humans, caterpillars often eat only certain parts of a plant, generally those that are the most palatable and digestible, and which have the highest food value. These include the buds and young leaves; flowers, fruit and seeds; starchy taproots and rhizomes. Sometimes the part of the plant that is eaten changes, often out of necessity, as the caterpillar grows or the season progresses. Two species found on pine, the Ochreous Pug *Eupithecia indigata* and the Pine Beauty *Panolis flammea*, feed entirely on the young male inflorescence, if available, when they first hatch in the spring, but switch to twigs and leaves respectively once it withers. A more remarkable example of sequential feeding, only recently discovered, is shown by *Acrolepiopsis marcidella*, probably the only moth in Britain that feeds on Butcher's-broom. First the caterpillar mines a cladode (false leaf, actually a flattened shoot), later it bores into the growing tip of a shoot, and finally completes its growth inside a berry (Sterling and Langmaid, 1998).

Less commonly, caterpillars eat parts of the plant that are clearly not the most nutritious. The clearwings and the Goat Moth *Cossus cossus* (Plate 6) bore into the living wood of trees and shrubs, and may take several years to reach maturity on this spartan diet. However, they are relatively safe from all but the most specialised predators such as woodpeckers and certain parasitic wasps. There are a few groups, like the fan-foots (Herminiinae), whose caterpillars prefer their leaves to be withered rather than fresh. Possibly the plant's chemical defences have begun to degrade in such leaves, or the habit may be a means of avoiding competition.

In their efforts to exploit every possible niche, some caterpillars have even gone underwater to feed on pondweeds. This is clearly a highly specialised lifestyle, confined at least in Britain to a few of the Pyralidae, notably the china-mark moths of the subfamily Nymphulinae. All inhabit still or slow-moving fresh water; there are no undersea moths.

Plate 6: A large head and powerful jaws enable the Goat Moth *Cossus cossus* caterpillar to bore into living hardwood. Its size and lurid colour sometimes attract attention as it wanders prior to pupation. (Norfolk).

Feeding on lichens and fungi

Feeding on lichens, although a minority habit, is surprisingly widespread among moths and involves a few members of many different families. It is common in the Tineidae, while among the macros, caterpillars of the Dotted Carpet *Alcis jubata* and the Brussels Lace *Cleorodes lichenaria* do it. Most of their nearest relatives feed on trees, so presumably it did not require much of a change of lifestyle to switch from leaves to lichens growing on the same twigs and branches. Two groups of macros, the footman moths (Lithosiinae) and noctuids of the subfamily Bryophilinae, specialise in feeding on lichens, including those growing on rocks as well as on trees. Although footman caterpillars are well camouflaged against a background of lichens, it is curious that none of the adult moths has adopted this policy. Indeed, it is hard to see the function of their plain, rather drab wing patterns (Plate 7). Are they trying to be cryptic, or are footman moths

Plate 7: Buff Footman *Eilema depressa*. Neither well-camouflaged nor with obvious warning coloration, moths of this genus perhaps mimic distasteful beetles. Two varieties of the female (dimorphism) are shown. (Norfolk).

unpalatable like most of their tiger moth family, the Arctiidae? To me, some resemble soldier beetles (Cantharidae).

Only one British macro feeds on fungi, the Waved Black *Parascotia fuliginaria*. This was once considered a moth of fabulous rarity, even though its life history was known. The caterpillars were found on bracket fungi growing on rotten wooden piles beside the Thames in London. Tutt (1902) mentions the difficulties of obtaining them. Since then, this moth has expanded its range considerably. Among the micros, the Nemapogoninae (a subfamily of the Tineidae) also specialise in feeding on bracket fungi. These fungi have fruiting bodies that persist for years, unlike the much more ephemeral and unpredictable mushrooms and toadstools.

Food of animal origin

For all their diversity, few Lepidoptera have managed to break away completely from a dependence on plants or fungi, and switch to food of animal origin, with all the new opportunities this can bring. As a family, again the Tineidae have been the most successful – as well as the most notorious, for they include the clothes-moths that can do so much damage to stored furs and fabrics. In the wild, they are found in birds' nests, owl pellets, mummified corpses, the dens and burrows of mammals, and anywhere there is detritus of animal origin. The Skin Moth *Monopsis laevigella* swarms around my own seldom mucked-out chicken shed, while up the hill its congener *M. weaverella* depends on fox droppings and the corpses of rabbits killed by myxomatosis. Unlike plants, such food is not seasonal, and in the right conditions some species can be continuous-brooded, especially inside caves and buildings. However, whilst they may have found a niche not occupied by other moths, it brings them into direct competition with those much more powerful, purpose-built insect scavengers, the beetles. This may explain why the habit is not more widespread.

Elsewhere in the world, there are some very unusual lifestyles. In North America, the caterpillar of the Planthopper Parasite Moth *Fulgoraecia exigua* (Epipyropidae) attaches itself to a frog-hopper (Homoptera), and lives as an ectoparasite, sucking body fluids from the abdomen (Covell, 1984). Far from being unique, a similar life history has also evolved in an unrelated Australian family, the Cyclotornidae, but with an added twist. After parasitising plant bugs (Hemiptera), the caterpillar becomes attractive to ants that are attending the bugs, and is carried back to their nest where it completes its growth on ant grubs (Watson and Whalley, 1975).

Here in Britain we have few or no entirely carnivorous caterpillars, but many that regularly supplement their diet by eating other caterpillars, especially when either their normal food or moisture is in short supply. Most notorious is the Dun-bar *Cosmia trapezina*, a quite harmless-looking caterpillar, but one that never hesitates to attack others even if they are larger than itself. The Small Quaker *Orthosia cruda* is another frequent cannibal, as is the Marbled Clover *Heliothis viriplaca*. I was once sent a packet of six of these caterpillars by first class post, but when it arrived there were only two and a third. One proved to be parasitised and became moribund, whereupon its companion ate both it and the wasp grub inside it. Such behaviour presumably removes rivals when food is scarce, as well as providing fluid and a high-protein meal for the aggressor. However, it is potentially risky – a bite from the loser could eventually kill the victor, or the prey might be ailing from a virus infection – this is why it seems to be a last resort for most species, and many never do it at all.

In Hawaii, caterpillars of several species in the pug moth genus *Eupithecia* catch and eat flies, striking with cobra-like speed. Perhaps this evolved from the habit, common to many caterpillars, of trying to bite any parasitic wasps and flies that molest them. Elsewhere in the Far East, other caterpillars in the same genus eat aphids. In Britain, the Ochreous Pug caterpillar can also supplement its diet with aphids (Leverton, 1998), and probably others will also be found that do this.

Growth and moult

Whatever its chosen food, the caterpillar grows until its skin will stretch no more. It stops feeding and becomes dormant, often anchoring itself on a pad of silk. Eventually, the old skin splits behind the head, and muscular contractions cause it to concertina downwards until it slips off completely. The technical term is ecdysis. Some caterpillars, such as the Sussex Emerald *Thalera fimbrialis*, then eat the cast-off skin, whether for its food value or to remove clues that predators might use. The head of a newly moulted caterpillar is disproportionately large, and the skin stretchable to allow for the next phase of growth. Often, the colour and pattern differ, sometimes strikingly so, from the previous stage.

The number of instars (stages between moults) varies according to species, and is not always constant, perhaps depending on speed of growth. For example, the Bordered White *Bupalus piniaria* has between four and six instars, according to conditions (D Barbour, personal communication). In macros, there are generally assumed to be five instars, but this is an over-simplification. Notable exceptions include the Zygaenidae, where the number of instars depends on how many times the caterpillar goes into a period of total inactivity known as diapause (Tremewan, 1985). I once reared the Forester Moth *Adscita statices* from the egg, and found it a most frustrating experience. The caterpillars grew normally at first, but after several moults they stopped eating and went into diapause. On reawakening they would moult again without feeding, so that they shrank rather than grew bigger. Some went into diapause more than once. It was difficult to keep count, but they were smaller in perhaps their eighth instar than they had been in their fifth! Some footman moth caterpillars moult even more frequently, with 10–14 instars recorded in the Hoary Footman *Eilema caniola* and 12 in the Four-dotted Footman *Cybosia mesomella* (Henwood, 1997). It is hard to see the advantage of this, as the time and resources required must be considerable.

Plate 8: The bright green sides that concealed the Puss Moth *Cerura vinula* caterpillar amongst foliage would be conspicuous against the bark where the cocoon is spun, hence they change to dull purple at this time. (Sussex).

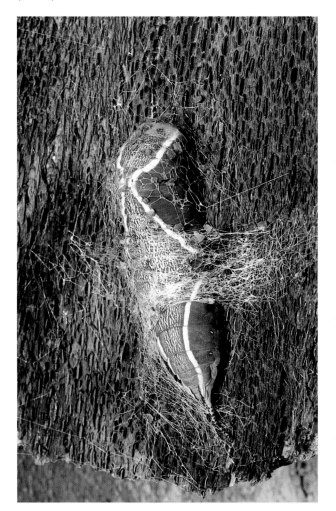

Preparing for pupation

For most moths, the caterpillar stage is the period of greatest mortality. Few reach full growth. Those that do, stop feeding, and rest until their last meal has passed through the gut. Sometimes there is a colour change: the mainly green caterpillars of the Kentish Glory and the Puss Moth *Cerura vinula* (Plate 8) take on a purplish tinge; in the Miller *Acronicta leporina* both the green body and its long white or yellowish hairs turn to a smoky grey. This change in appearance is associated with a behavioural one, as the caterpillars leave the green foliage and begin to wander, searching for a suitable pupation site. It is absent in species such as the Clouded Drab, where the caterpillars descend the tree at night, and immediately tunnel into the soil.

However, for many species a period of wandering before pupation is obligatory. Many naturalists must have seen the hairy brown caterpillar of the White Ermine *Spilosoma lubricipeda* frantically scurrying across a busy road, and perhaps rescued it. Linnaeus's choice of specific name, which translates as swift footed, was memorably apt. One purpose of the behaviour may be to avoid concentrations of pupae close to the plants where the caterpillars feed, the natural place for predators to search. In a mixed Speyside wood, Mark Young and I found more Puss Moth cocoons on the trunks of Silver Birches than on the Aspens where the caterpillars had fed.

The pupa

The final time that a caterpillar moults it becomes a pupa. Whereas in butterflies this stage is usually exposed to view, in moths it is generally hidden away. The most common sites, at least for the macros, are inside a slight cocoon of silk spun up between dead or living leaves, amongst ground debris, under loose bark, or below the surface of the soil. Caterpillars that live in borings often pupate there too. A few groups make more substantial cocoons, like the pale yellow and highly visible ones of the Six-spot Burnet *Zygaena filipendulae*, the complicated exit-only construction of the Emperor Moth, and the hardened case of the Puss Moth (Plate 97). Some caterpillars do not spin a cocoon as such, but merely shape out a cell beneath the soil or in detritus, like the Pine Hawk-moth *Hyloicus pinastri* and the Northern Dart.

The pupa is the most vulnerable stage: usually pupae can wriggle, but have no means of escape if attacked. A few are protected by an unpleasant taste, while those of the tussock moths are hairy. If the caterpillar is armed with irritating hairs these are woven into the cocoon, and serve equally well to protect the pupa. However, most pupae depend for their survival mainly on the caterpillar's selection of a suitable pupation site.

Development of the imago

The next act of metamorphosis may start immediately, or after a period of diapause, which in a few species can last for several years. Inside the pupa, the heart and respiratory system remain, but many of the other tissues break down into a kind of soup, and begin to reform into the wings, eyes, legs, antennae and all the other parts of the future moth. If there is no diapause, for most British macros it takes between 3 and 6 weeks after pupation for the moth to become fully formed and ready to emerge, depending on the species. Cold retards the process, and warm weather speeds it up.

Emergence

Most species have a particular time of the day for eclosion (emergence from the pupa). Day-flying moths usually emerge in the morning, fairly early in

Plate 9: As this female Kentish Glory *Endromis versicolora* shows (left), the unexpanded wings are not crumpled or creased on emergence from the pupa. Their scales are densely packed, standing upright on the wing surface. Once the wings are expanded (right), all moths hold them closed over the back in the 'butterfly position' until they are dry. The scales now lie flat, overlapping slightly like roofing tiles. (Moray).

the case of clearwings, but between mid and late morning for burnets, and the Oak Eggar and Emperor moths. Surprisingly, many wholly nocturnal moths also emerge in broad daylight, though others wait until dusk or after dark. It is a mystery how those that pupate deep in the soil or in other dark places manage to time the process correctly. Perhaps the rhythm of daily rises and falls in temperature plays a part.

Emergence is clearly quite a strenuous activity. First the moth must crack open the pupal shell, emit fluid to soften the cocoon if there is one, then force its way out. If the pupation site is underground, the moth must tunnel its way to the surface. Normally, it then begins to climb, until it reaches a suitable place with secure footholds on a vertical or overhanging surface. Most moths then rest for a few minutes before inflating their wings. Having done this, they are still unable to fly until their wings have dried and stiffened, thus newly emerged moths are particularly vulnerable. For macros, the whole process generally takes between 1 and 2 hours, but again warmth speeds it up.

A frequent misapprehension is that the wings of butterflies and moths are 'crumpled' when they emerge from the pupa. They are not (Plate 9) – unless the individual is deformed. The appearance of crumpling is caused by the way the wings are inflated. A fluid (haemolymph, the moth's equivalent of blood) is pumped into the veins of the wing, extending them and in turn stretching the wing membrane. Inflation begins from the basal area of the wing and gradually proceeds towards its tip. Until the process is complete, buckling occurs at the boundary between the inflated and uninflated areas of the wing. Both in captivity and in the wild, any serious hitch in this process can leave the unfortunate moth a cripple, unable to fly. A male so handicapped would be unlikely to reach a mate, but a female would be less disadvantaged. The females of some species are flightless anyway, but very few species are known where the male is also flightless, except on some small and remote oceanic islands.

Mating

In day-flying species, the female begins to 'call' by releasing pheromones (complex and highly volatile chemicals that attract the males) as soon as, or even before, she has dried her wings. Nocturnal moths that emerge from the pupa during the day wait until dusk or after dark before doing this. Generally the female mates before she flies, except for those species that hibernate and mate in the spring, and those that migrate with undeveloped ovaries to save weight. However, my captive-bred Buff-tip females flew at dusk without calling, and when caged with males did not mate until the third night. As the caterpillars feed communally, this may have been a means of reducing the danger of inbreeding through mating with their own siblings.

There is little courtship in moths. The pheromones used by the female to attract a mate, and sometimes by the male to stimulate the female after he arrives, are nearly always unique. Thus there is no need for any further recognition behaviour to establish that the suitor belongs to the same species and is of the correct sex. The right perfume is everything, at least for nocturnal moths! Nor is there much evidence that the female ever rejects the male that reaches her first – perhaps the very fact that he has done so is sufficient proof of his quality. If several males reach a female at the same time, jostling takes place, and the most vigorous and capable male does seem to win. I have seen this happen in the wild when four male Drinker Moths *Euthrix potatoria* attended a female that was awkwardly positioned in long grass. The smallest, rather yellowish male attempted to mate, but was barged out of the way by a much larger brown one. However, under laboratory conditions at least, Scarlet Tiger *Callimorpha dominula* females do show a preference for a male of a different genotype from themselves (Kettlewell, 1973), but the way they express this was not stated.

Males can mate with more than one female. Two virgin female Emperor Moths put out in my garden attracted only a single wild male. After pairing with one of the females he flew off, but immediately circled back and paired with the second female. Both laid fertile eggs. In captivity, a male Vapourer paired with three different females, and all the subsequent egg batches hatched. Although the sex ratio in moths probably approaches equality, this ability of males to mate several times has the effect of increasing the competition for females, which generally mate only once.

The length of time that moths stay paired varies. In the Emperor Moth and Oak Eggar the sexes separate after about 20 minutes to half an hour, but in the Kentish Glory, similar in size and habits, moths stay paired for 2 or 3 hours (Plate 82). Possibly some element of mate guarding is involved (preventing the female from pairing again with a different male). However, I once accidentally disturbed a pair that had been together for about half an hour, causing them to separate. The female immediately began to call again, suggesting that the male's packet of sperm had not yet been transferred, and the two soon re-mated. Some moths that mate at night stay together (*in copula* is the scientific term) all through the following day. These are usually fairly large species, or as in the case of tiger moths, warningly coloured. I have seen this behaviour in the Poplar Hawk-moth *Laothoe populi*, Bufftip, Puss Moth, Dark Tussock, White Ermine and Sword-grass (Plate 101), among others.

The number of broods (voltinism)

In warmer countries, the life cycle from egg to moth can be completed very quickly. In the United States, the Fall Armyworm Moth *Spodoptera frugiperda* has up to ten broods a year in the Deep South, where it is a major pest of crops such as maize, cotton, grain, tobacco and many vegetables. These caterpillars can cause so much devastation that they are forced to migrate *en masse* to seek fresh food supplies, hence the common name (Covell, 1984). Even in Sussex, I once reared a brood of a related species, the Small Mottled Willow *S. exigua* (known as the Beet Armyworm in North America) from egg to adult (Plate 10) in only 48 days. It too is a pest of crops in warmer regions, but only a scarce migrant to Britain.

In temperate countries, winter limits the number of generations possible in a year. Even those continuous-brooded micros that inhabit warehouses and other heated buildings and feed on stored products do not usually have more than three broods. In the

Plate 10: Small Mottled Willow *Spodoptera exigua*. Members of this genus are notorious agricultural pests in warmer regions, but cannot survive in the wild in Britain. (Bred from an immigrant female, Sussex).

wild, the great majority of Britain's moths are univoltine (single-brooded), even in the south, and the percentage increases northwards. Sufficient information is not available for all the micros, but of about 570 macro-moths that can be considered permanently resident in Sussex, 88 (just over 15 per cent) normally have more than one brood a year, but in many cases this is only partial (data extracted from Pratt, 1981, and Emmet, 1992). A handful of these species occasionally produce a third brood in good years. However, when broods overlap, or extend over a period of many weeks, it is not easy to be sure how many there really are. Even for some of our commonest moths, solid evidence is lacking.

Of course, it is not only climate that prevents moths from having more than one brood. Many are limited by their food plant. The caterpillar of the Foxglove Pug *Eupithecia pulchellata* feeds on stamens in the flowers, that of the Green Pug *Chloroclystis rectangulata* feeds on apple blossom, others need fresh young leaves, or unripe seeds, all of which are seasonal. Unless they can use an alternative food plant for the second brood, like the Holly Blue butterfly *Celastrina argiolus*, such species are limited to a single brood.

Many species that are double-brooded (bivoltine) in the south become univoltine at higher latitudes, the dividing zone in Britain often being in northern England. In Banffshire, nearly 500 miles north of Sussex, only 2 (0.66 per cent) out of 301 resident macro-moths are regularly double-brooded. These are the Grey Pine Carpet *Thera obeliscata* and the Garden Carpet *Xanthorhoe fluctuata*. Further north, in the very short Shetland summer, the latter species becomes univoltine too (M Pennington, personal communication). The simplest explanation would be that the moths are unable to fit in a second brood because the cooler temperatures slow down the development rates of the early stages, coupled with a summer that arrives later and ends sooner. If this were the case, one might expect second broods of potentially bivoltine species to appear during exceptionally long and warm northern summers, but this does not happen to any great extent. Moreover, even when stock of northern origin is transported further south, it often continues to be univoltine. This suggests that a genetic element is involved. Perhaps it ensures that a species adjusted to an average cool summer is not fooled by an unusually warm one, so does not attempt to produce a second brood that will have little chance of survival. However, I suspect that there are positive benefits in having only one brood where the climate is uncertain.

Insuring against poor years

Often, the single northern brood emerges over a period almost as long as that taken by the two generations in the south. In north-east Scotland, the Sallow Kitten *Furcula furcula* (Plate 11), Iron Prominent *Notodonta*

Plate 11: Sallow Kitten *Furcula furcula*. The long tails of this miniature dragon are greatly modified anal claspers enclosing red flagella that can be everted and waved, perhaps as a chemical defence against other invertebrates. The univoltine northern population favours low moorland sallows. (Banffshire).

dromedarius, Pebble Prominent *N. ziczac* and Lesser Swallow Prominent *Pheosia gnoma* fly from mid or late May to early August – by which time their second broods would be out further south. Hedging bets is the most likely advantage. Even in the worst years, it is unlikely that the weather will be adverse throughout this time, so at least part of the brood should encounter reasonable conditions. Eggs and caterpillars too will be spread over a much more extended period, again minimising the risk that all will be at the same vulnerable stage during a spell of bad weather.

Another means of surviving occasional disasters is the habit, common in many Scottish species that overwinter as a pupa, of missing out a year. It is also found in moths dependent on transient weeds of disturbed ground, like the Mullein Shark *Shargacucullia verbasci* and the Grey Carpet *Lithostege griseata*. Instead of emerging the following summer, they remain as pupae for a further 12 months or even longer, sometimes for 4 or 5 years. This would make good sense if the pupa somehow 'knew' it was going to be a bad year for the food plant or the weather, but there is little evidence for that. Such a feat would be remarkable in those species where the moth is fully formed inside the pupa by autumn if the moth is to emerge the following year. The pupa would need to be able to predict conditions 7 or 8 months in advance – far beyond human capabilities. Even in those species where the moth does not form up until the spring, the forecast would have to be correct for at least 1 month ahead.

The habit is so widespread that the advantages must be sufficient to outweigh the obvious dangers: instead of having to survive perhaps 8 months, the pupa is exposed to the risk of predation, disease and misadventure for 20 months, 32 months or whatever. Nor is there any guarantee that the year when it does emerge will be any better than the ones it skipped. However, it may be that if the pupation site was sufficiently well-chosen for the pupa to survive its first winter, the risk of subsequent mortality is low. When the strategy does work successfully, the benefits to the lucky individuals will be considerable, enabling them to repopulate the whole area with their descendants.

Overwintering strategies

In all temperate regions, moths must find a way of passing the winter safely. The particular stage of the life cycle used for this is fixed in nearly all species. The choices are as follows:

1. As an egg
   - Undeveloped
   - With the caterpillar already fully formed inside the eggshell
2. As a caterpillar
   - Small, or partly grown, but in complete diapause until spring
   - Small, or partly grown, feeding during mild weather
   - Fully grown, having finished feeding
3. As a pupa
   - Undeveloped
   - With the moth already fully formed inside the pupa
4. As a moth
   - Active, weather permitting – the normal season of emergence for some species
   - In complete diapause (hibernation) until the spring
   - In diapause, but occasionally active during mild spells.

Young (1997) gives a table showing the percentages of resident British moths that overwinter in each of the four main stages, classified by families. Unfortunately, this information is not available for a surprisingly large number of microlepidoptera and this will have distorted the picture to some extent. However, Young's table does suggest consistent differences between micro- and macro-moths, with the former less likely to overwinter as an egg (2.3 per cent versus 16 per cent) and more likely to do so as an imago (9.1 per cent versus 2.8 per cent).

Passing the winter as an egg

A curious fact that does not seem to have been pointed out before, is that all resident British moths and butterflies that overwinter as an egg (about 158 species) are invariably single-brooded. Even in captivity, it seems that none can be persuaded to produce another brood. The Vapourer, repeatedly credited with a second brood, might be thought an exception. However, West (1993) correctly queried this, suggesting that moths are seen on the wing from July to October because of the staggered hatching of eggs. Earlier authors such as Tutt (1902) had already described how those in the same batch sometimes hatched over a period of 5 or 6 months. This confusion demonstrates how we often know more about scarce and local species than the really common ones that nobody bothers to study – unless these are commercial pests.

One such pest is the Antler Moth *Cerapteryx graminis*. Over most of its range, it is a harmless and inoffensive grass-eating species, but in certain parts of northern England and southern Scotland its numbers periodically increase so much that the caterpillars cause severe damage to upland pastures (South, 1907–09). The eggs are dropped in flight, and overwinter with the caterpillar fully formed inside the shell. Goulson and Entwhistle (1995) obtained 10,000 eggs from 300 females captured by two light traps in one night near Moffat in the Scottish borders (giving some indication of its abundance). The eggs were divided into batches, and subjected to different regimes of heating, cooling and humidity, but all attempts to make them hatch before they would have done in the wild were unsuccessful.

This may explain why no species that overwinters as an egg can have more than one brood a year. Presumably, the mechanism that prevents eggs hatching prematurely during warm spells in autumn and late winter has to be so foolproof that it does not have the flexibility to allow them to hatch quickly at a different time of year. Another feature of such eggs is their unusually tough and thick shell, again hinting at the degree of specialisation this habit requires.

Passing the winter as a caterpillar

This is the option chosen by most British moths, but it is very much influenced by food plant. Many herbs and grasses stay green even in the severest winters, while tussocks and basal rosettes provide cover and shelter, enabling caterpillars to nibble a little during mild intervals. Thus whole groups that feed on grasses, like most of the subfamily Crambinae and noctuids in the genera *Mythimna*, *Apamea*, *Oligia* and *Photedes*, overwinter as partly grown caterpillars. As do many that are polyphagous on low plants: for example most of the waves (Sterrhinae), many of the tiger moths (Arctiinae) and most of the darts (Noctuinae). Those that are found on lichens, like the footman moths, can also continue to feed through the winter.

However, there is one lifestyle that is largely incompatible with overwintering as a caterpillar and that is feeding on the leaves of deciduous trees. In autumn, both food and shelter disappear for many months. Therefore, most of the moths in this category pass the winter in one of the other stages. Leafminers in the family Nepticulidae almost all overwinter as a pupa; those in the Gracillariidae do so as a pupa or as an adult; many noctuids as an egg. Groups that do overwinter as a caterpillar on trees are forced to make special arrangements. The Coleophoridae live inside a silk case constructed inside a fragment of the food plant, and so are protected. The tussock moths spin a web for hibernation. Among the Geometridae, the Large Emerald *Geometra papilionaria* changes from green like the birch leaves to brown like the bare twigs, and anchors itself on a pad of silk. Other geometrids such as the Willow Beauty *Peribatodes rhomboidaria* have caterpillars that closely resemble twigs anyway, and are not dependent on green foliage for camouflage. All these species have to go through the winter without food, however, until the buds start to burst in the spring, so this is true hibernation.

Perhaps this explains why, of about 72 British species in the large family Noctuidae that are entirely dependent on deciduous trees, only one overwinters as a caterpillar. This is the Scarce Silver-lines *Bena bicolorana*, which feeds on oak (Plate 12). Like the caterpillar of the Large Emerald, it turns from green to brown in autumn, and takes up a position near a young leaf bud, which it much resembles (Heath and Emmet, 1983). (Ironically, there has always been some dispute as to whether the small subfamily Chloephorinae to which it

Plate 12: The Scarce Silver-lines *Bena bicolorana* is one of few bright green moths. Its forewings have a simple disruptive pattern. The proper place of its group in the system of classification has long been debated. (Sussex).

belongs is part of the Noctuidae anyway. The ancients regarded the moths as tortricids, hence the specific name ending with *-ana*, while recently Karsholt and Razowski (1996) have transferred them to the Nolidae.) Other noctuids get round the problem of leaf fall by feeding on low plants until the spring. Then, when the fresh new foliage appears, the caterpillars climb the trees to take advantage of it. Noted exponents are the Triple-spotted Clay *Xestia ditrapezium*, Grey Arches *Polia nebulosa*, and Old Lady *Mormo maura*, but there are many others.

Passing the winter as a pupa

Passing the winter as a pupa seems to bring with it no constraints. This strategy is used by a wide variety of species having different food plants and lifestyles, whether or not they have more than one brood a year. However, if the moth is already fully formed within the pupa by the autumn, that species is always univoltine; at least, I am not aware of any exceptions. Species that use this tactic come from a wide range of families, but nearly all fly in early spring, like the Small Eggar *Eriogaster lanestris*, Yellow Horned *Achyla flavicornis*, Rannoch Brindled Beauty *Lycia lapponaria*, all the Quaker moths in the genus *Orthosia*, and the Rannoch Sprawler *Brachionycha nubeculosa*. They are able to respond very quickly as soon as the weather becomes mild enough for emergence, without having to wait several weeks for the imago to develop inside the pupa. In effect, the moth itself is hibernating within the pupal case. There is a strong tendency in this group for pupae to skip a year, and overwinter more than once, as already described. If this is going to happen, the pupa passes the winter in an undeveloped state.

Passing the winter as an adult

Winter is the normal flight time for a few of the Geometridae. Perhaps these should be regarded as not having a hibernating phase. All have flightless females (Plate 13), sometimes described as wingless, but as the wings are reduced to stubs rather than absent (apterous), the correct term is brachypterous. Curiously, the female pupae have full-sized wing cases, just as in the male, but they remain empty when the moth forms up inside. In contrast, in the Vapourer the wing cases of the female pupa are as rudimentary as the wings. Possibly the loss of functioning wings is (in evolutionary terms) a recent development in the Geometridae. In both the Winter Moth *Operophtera brumata* and Northern Winter Moth *O. fagata* the newly emerged female still holds the remnants over her back as if drying them in the normal way (Plate 79). Perhaps in the distant past these species had two forms of female, one fully winged and one flightless, like the strange aquatic pyralid, *Acentria ephemerella*.

Moths that fly in the middle of winter obviously must be able to withstand low temperatures. I once found a male Pale Brindled Beauty *Apocheima pilosaria* that had nose-dived into a pool of meltwater. This had refrozen, and the moth was almost completely encased in ice 1.5 cm thick. Having chipped out the piece containing the moth, I took it home and thawed it out. As expected, the moth flew away at dusk that evening.

Taken overall, hibernation as an adult moth is the least common strategy, but one used by a very wide range of families. Some of these moths look far too delicate and fragile to survive the winter – the *Caloptilia* species, plume moths like *Amblyptilia acanthadactyla* and *Emmelina monodactyla* – but appearance is deceptive. Among the macros, noctuids presently placed in the subfamily Cuculliinae are the main exponents. They can be found building up their resources in autumn by sucking the juice of overripe berries, or attending the lepidopterist's sugar patch, reappearing in the spring at sallow blossom. The Brindled Ochre *Dasypolia templi* is atypical in that the adult moth cannot feed, but relies on reserves built up by the caterpillar.

The actual sites used for hibernation are known for only a few species. The Herald

Plate 13: Spring Usher *Agriopis leucophaearia*. Amply winged males but flightless females are characteristic of geometrids that emerge in winter or early spring – February in this case. (Norfolk).

Scoliopteryx libatrix and the Tissue *Triphosa dubitata* regularly use caves, or their man-made equivalent, tunnels and empty buildings such as concrete wartime bunkers (Morris and Collins, 1991). As they are found in areas where these are not available, they must use other sites too. The Oak Nycteoline *Nycteola revayana* is sometimes beaten from Yew. Twice I have found the Tawny Pinion *Lithophane semibrunnea* hibernating on the underside of wooden rafters in old farm outbuildings; the nearest equivalent in the wild would be beneath a large bough. *Agonopterix heracliana* can sometimes be found under loose bark, and *A. umbellana* amongst the mass of dead needles accumulating in the heart of old gorse bushes.

In a few species, hibernation is only partial. Even in north-east Scotland, the Chestnut *Conistra vaccinii* and the Satellite *Eupsilia transversa* (Plate 14) will come to sugar on mild January nights. By early March, both may be quite worn, unlike the true hibernators that often reappear in spring looking almost as fresh as they did in autumn. These adult hibernators are our moths with the longest life span. Most emerge in late September in southern Britain (but earlier in the north) and some survive into May, or occasionally June in Scotland. Even allowing for the time spent in diapause, and their often fairly sluggish behaviour before and after, they have an active life of at least 2 or 3 months, compared with the week or two of most other species. In three British species, the Brindled Ochre, Red-green Carpet *Chloroclysta siterata* and Autumn Green Carpet *C. miata*, mating takes place in autumn, whereupon all the males die and only the females hibernate. Sexual equality is not yet a feature of the moth community.

Plate 14: Satellite *Eupsilia transversa*. When fresh, the rich russet shades camouflage this moth amongst fallen autumn leaves, the white 'moon' perhaps representing a hole. Hibernation is only partial, so by spring moths are often bleached and worn like the leaves themselves. (Banffshire).

Chapter 2

Colours, patterns and shapes

We know the main causes of mortality in moths, but not their relative importance to population dynamics. Losses caused by adverse weather, food shortage, misadventure, parasitism and disease probably outweigh those due to predation, but these losses would still occur whatever the moth looked like. Predation is different: it is affected by the outward appearance of the moth. Both in the early stages and as adults, moths have evolved either remarkably cryptic shapes and patterns to frustrate visually searching predators, or warning coloration combined with other defensive features to deter them.

The imago

Possession of scaly wings is the defining characteristic of the Order Lepidoptera. The wing membranes themselves are colourless and transparent. Only the scales are pigmented, or structured so as to refract coloured light. Individual scales are all one shade, but being so small they blend in the eye of the viewer to produce subtle gradations and tones, like a canvas painted using the technique of pointillism. This has enabled butterflies and moths to develop more varied and complex wing markings than any other insects. Because both surfaces of the wing are scaled, it is possible for the upper and lower surfaces to be very different in appearance, and this is usually the case. In butterflies, the upper surface is often eye-catching – being brightly coloured and strongly patterned – whereas the underside provides a subtler camouflage when the wings are closed over the back at rest. The Peacock *Inachis io* and Small Tortoiseshell *Aglais urticae* are obvious examples. However, nearly all moths rest with the upper surface of the wings exposed, so this surface is cryptically coloured except in warningly coloured species. In effect, the upperside of a moth's forewing is equivalent to the underside of a butterfly's hindwing. Wing surfaces that are not in view during the daytime tend to be weakly marked or almost plain, thus moths in a collection are rarely mounted upside down to show the underside. This has a dedicated pattern only when the resting posture leaves it visible, as in the Thorns *Selenia* and *Ennomos* (Plates 15, 86), or in day-flying species, including those using flash coloration if disturbed (Plate 31).

To avoid being eaten, most moths adopt camouflage as their main line of defence. This is clearly the safest option providing it is successful: a predator cannot attack prey unless it finds it first. Unfortunately, if the camouflage fails the consequences tend to be fatal, so it needs to be good. There are countless breathtaking examples of moths that have evolved an almost perfect match to the background – bark, rocks, vegetation – on which they habitually rest. Or there is the opposite tactic: bright primary colours juxtaposed in sharp, simple patterns to make the moth conspicuous and at the same time convey a warning message to would-be predators. This is adopted by a minority of species overall, but by some whole families. Usually such moths contain distasteful, even toxic, substances, or mimic others that do. This strategy also entails a risk, in that a specialist predator might be immune to the moth's chemical or physical defences, or an inexperienced one could injure the moth before rejecting it.

Plate 15: Unlike most other moths, the Early Thorn *Selenia dentaria* rests like a butterfly. Hence its underside, being in view, is fully marked and patterned – mimicking a withered leaf. (Banffshire).

In practice the distinction between cryptic and aposematic coloration is not clear cut. Some species combine the two, like the Eyed Hawk-moth *Smerinthus ocellata*: if its disguise is penetrated the cryptic forewings are raised to reveal menacing eye-spots on the hindwings. Other moths are not obviously warningly coloured, yet seem not particularly well camouflaged either. If the Yellow Shell *Camptogramma bilineata* (Plate 84) emerged in September we might suppose it was meant to resemble an autumn leaf, but it flies in July. The Smoky Wave *Scopula ternata* is not at all well hidden when it sits on its food plant, heather – unlike its close relative the Lewes Wave *S. immorata* in similar circumstances. However, their appearance has not evolved for human eyes. To a predator these moths might be well disguised, or possibly aposematic enough not to be worth the trouble of catching unless other food is scarce. There is no sharp divide: moths present a gradual spectrum of unpalatability ranging from the lethally toxic to the mildly unpleasant, and even cryptic species can contain distasteful substances.

Sexual recognition

While reducing predation seems to be the main purpose of wing pattern in moths, we should not assume it is the only one. In some species it might have a sexual or a physiological function.

Plate 16: Narrow-bordered Bee Hawk-moth *Hemaris tityus*. This diurnal moth mimics a bumble bee, but in flight its greater speed and agility while hovering to feed at flowers can be recognised with practice. (Moray).

Butterflies initially find mates by sight, as proved by their mistakes. Male Small Whites *Pieris rapae* often investigate white flowers, male Meadow Browns *Maniola jurtina* check out every withered leaf lying in the turf. Once a real female is located, courtship displays and the exchange of pheromones confirm compatibility. Rival males are also recognised, and in territorial species airborne battles for supremacy often ensue. Wing patterns therefore have a sexual function in this group. (It is hard to explain why humans so often consider these patterns beautiful too.)

In contrast, there is little evidence that moths use visual clues for the same purposes. This would be difficult for truly nocturnal species, but even

diurnal ones apparently locate their mates by scent, with sight playing little if any part. Nor are any male moths known to hold territories, at least in Britain. Thus aerial attacks on rivals and courtship chases, so familiar in butterflies, are almost absent. One exception is the Narrow-bordered Bee Hawk-moth *Hemaris tityus* (Plate 16), where male and female engage in rapid pursuits low over the ground, or spiral upwards together (Pittaway, 1993). Similar chasing is also recorded for the Humming-bird Hawk-moth *Macroglossum stellatarum*, and twice I have seen males returning to perch in the same spot after aerial forays. In these species, the abdominal markings and anal tufts presumably aid visual recognition in the same way as tail patterns in finches do.

Other such clear-cut instances are unknown among British moths, but there may be a few species where sight plays some part during mating. Hovering male Ghost Moths *Hepialus humuli* and Gold Swifts *H. hecta* emit a pheromone to attract females (a reversal of the usual roles), but their conspicuous wing colours – silvery white, and with metallic blotches respectively (Plate 1) – presumably help females to locate them. It is also probable that males of many species, once attracted into the vicinity of a female by her scent, make the final contact visually. Both Oak Eggar and Emperor males appear to do so.

Heat absorption

Black objects absorb more of the sun's radiation than white ones, hence they warm up more rapidly. There is some evidence that Lepidoptera in colder regions, either high altitudes or high latitudes, are blacker than those elsewhere. Kettlewell (1973) describes this as 'Thermal Melanism'. Among butterflies, the northern and alpine genus *Erebia* is often given as an example, all its members being dark brown. However, the Ringlet *Aphantopus hyperantus*, a lowland species, is blacker than any of them. The Large Heath *Coenonympha tullia* has a very northern distribution, yet is relatively pale for a member of the Satyrinae, to which all these species belong.

Nevertheless, day-flying upland moths in Britain do tend to have much black in their coloration. The newly re-discovered micro *Ethmia pyrausta* is one example. Both the Black Mountain Moth *Glacies coracina* and the Netted Mountain Moth *Macaria carbonaria* (Plate 68) are dark, but with considerable whitish freckling. The Small Dark Yellow Underwing *Anarta cordigera* has an intensely velvet-black body and forewing, albeit dusted with silvery white. I have certainly seen it warming up in the sun prior to flight with the abdomen partly exposed. On the other hand, two mountain pyralids that often fly by day, *Udea decrepitalis* and *U. uliginosalis*, are relatively pale species, and our blackest moth, the Chimney Sweeper (Plate 28), is neither northern nor upland. As with butterflies, the evidence for thermal melanism is not overwhelming, although it may be one element that influences the pigmentation of Lepidoptera.

Cryptic coloration

The great majority of moths adopt crypsis as their defence against predation. It would be difficult to give an exact percentage, especially if the micros are included, because whole groups cannot easily be categorised. For instance, many of the numerous tiny *Stigmella* and *Phyllonorycter* species have striking and often metallic markings on their forewings, but their function is unclear. These moths are so small that perhaps they are beneath the notice of most vertebrate predators. For the macros, and the larger micros, it is usually possible to work out (or guess) the reason for their coloration with varying degrees of confidence, especially when the moth's behaviour and resting site is taken into account.

The most highly perfected examples of camouflage are seen in moths that rest openly, rather than supplement their cryptic appearance by hiding out of direct view. Those that habitually sit on the more or less vertical surfaces of tree trunks or rocks, or rest on twigs and foliage some distance above the ground, usually resemble their background to an astonishing degree. By selecting such resting places, these moths narrow but intensify their exposure to predators. They are virtually safe from small mammals such as shrews, voles and mice, from reptiles and amphibians (at least in Britain) and from carnivorous ground invertebrates such as beetles. On the other hand, they are particularly exposed to birds. The vision of birds, including their perception of colour, is much like our own, except that they are more sensitive to ultraviolet wavelengths. Also, their colour discrimination is more acute. Hence it is not surprising that the shapes, patterns and resting postures that have been perfected over millennia to frustrate avian predators are equally convincing to the human eye.

There are constraints to the structural adaptations involved in such camouflage: the shape of the wings can be modified to some extent, as in the hook-tips (Drepanidae) (Plate 17), but not so outlandishly as to interfere with their main function, flight. The leading edge (costa) of the forewing needs to be mechanically strong and relatively straight for most of its length. *Acleris emargana* is unique

Plate 17: Pebble Hook-tip *Drepana falcataria*. Most moths of this family have falcate tips to their forewings, and may benefit from being unlike the usual 'moth-shape' more familiar to searching predators. (Sussex).

Plate 18: Mottled Beauty *Alcis repandata*. Resemblance to tree bark is perfected by this common species. It varies regionally to suit its background, grey forms that rest on rocks being found in Scotland. (Banffshire).

among British moths in having the costa deeply scooped or emarginate. Splits or holes in the wing, or an uneven surface, must be suggested by markings. Long tufts of hair scales on the thorax or abdomen, or exaggerated palps as in the case of the Pale Prominent *Pterostoma palpina*, can superficially alter the shape of the moth without requiring major structural changes. An unusual resting posture can likewise disguise its basic appearance to a remarkable degree.

Cryptic camouflage can be divided into two main types, but with much overlap. In the first, the moth seeks to blend so well with its background that it is totally indistinguishable from it. Species that resemble tree bark, like the Mottled Beauty *Alcis repandata* (Plate 18) and its allies, fall into this category. So do those that rest on rocks, like the Glaucous Shears *Papestra biren* (Plate 69) and the Grey Mountain Carpet *Entephria caesiata*. There is conflicting evidence as to whether such moths deliberately choose appropriate resting sites to match their coloration, or merely rely on the odds being in their favour within their chosen habitat. After all, bark and rocks are far from uniform, being a mosaic of lighter and darker patches. Moths in this group tend to be variable, a subject that will be discussed later. They also frequently evolve local forms adapted to the predominant rock type or tree bark, and its lichen cover or lack of it.

The other strategy is to resemble a common inedible object of no interest to a predator. In these cases, the moth is seen, and distinguished from its background, but not recognised for what it is. Such species tend to be very constant in appearance with little variation. Having perfected their resemblance, presumably there is no advantage in departing from it. Withered leaves are frequently imitated, and both the Angle Shades *Phlogophora meticulosa* (Plate 117) and Lilac Beauty *Apeira syringaria* (Plate 19) have pleated forewings to enhance the deception. Forewings of 'dead leaf' moths often have a white or silvery central marking apparently representing light shining through a hole or crack, analogous to the feature on the underside of the hindwing that gives the Comma butterfly *Polygonia c-album* its name. In the Beautiful Golden Y *Autographa pulchrina* the mark is similarly formed by metallic scales, but in the Lunar Thorn *Selenia lunularia* (Plate 107) the central crescent on all four wings is genuinely translucent because it lacks scales. Rather few moths resemble green leaves, notably the emeralds (Geometrinae) (Plate 106), but numerous autumn moths adopt the bright yellows, oranges and russets of

Plate 19: Bent and folded like a piece of origami, the Lilac Beauty *Apeira syringaria* excessively resembles a yellowing leaf. Even so, it is less numerous than related species whose camouflage seems (to us) far more perfunctory. (Norfolk).

Plate 20: The Buff-tip *Phalera bucephala* takes camouflage to extremes, mimicking a snapped-off piece of dead twig with flaking bark and one jagged end. Although highly cryptic, it may be distasteful too, like its caterpillar (Plate 115). (Banffshire).

turning and fallen ones, complete with markings suggesting scars, mould and decay (Plates 14, 76, 86). Among geometrids, the Ennominae provide many such examples but the Larentiinae none, which is surprising since some of its members do fly in autumn.

Other species resemble a fragment of dead twig or stalk, even down to its brittle, broken ends. The Buff-tip (Plate 20) is the most elaborate in its deception, but the Flame *Axylia putris* and the Sword-grass *Xylena exsoleta* (Plate 101) are also convincing. In these last two, the generic name remarks on their resemblance to wood – xulon in the Greek – while the specific name adds that this is old and rotten (Emmet, 1991).

Moths confined to one main food plant, on which they normally rest, form an interesting sub-group. They have the opportunity to develop a precise resemblance to it, rather than a more generalised camouflage. Usually they are species whose food plant occurs in large dominant stands, where there is a good chance that the moth will settle on it even when landing at random. Many unrelated species associated with heather moorland have an intricate pattern often mingling red-brown, white and dark brown, making them well concealed at rest. They include the tortricid *Olethreutes schultziana*, the Common Heath *Ematurga atomaria* and the True-lover's Knot *Lycophotia porphyrea*. In grassland and fens, assorted noctuids collectively known as wainscots have evolved a similar straw-coloured appearance (Plate 33), as have many of the Crambinae.

It is harder to be sure that the camouflage is specific when the food plant is a tree – most bark looks much alike, except for that of the Silver Birch. Many moths associated with this tree are indeed white or silvery grey like its bark, including the Birch Mocha *Cyclophora albipunctata*, the

Plate 21: Pattern and resting posture combine to conceal the Streak *Chesias legatella* on the branches of its food plant, Broom. Note how the left forewing moulds to the contour of the branch. (Banffshire).

Lesser Swallow Prominent, the Miller, and the micro *Acleris logiana* in the Scottish Highlands. To be different, the Scalloped Hook-tip *Falcaria lacertinaria* instead resembles a curled dead birch leaf. The younger bark of the Scots Pine is reddish, and is copied by several moths that rest on its trunk, such as the Pine Carpet *Thera firmata* (Plate 66) and several tortricids, especially the Pine Shoot Moth *Rhyacionia buoliana*.

Otherwise, there are relatively few convincing instances in British moths. Before it became extinct here, the Frosted Yellow *Isturgia limbaria* was one example, sitting like a butterfly with the streaked underside of the hindwings matching the shoots of its food plant, Broom (Ford, 1955). The Streak *Chesias legatella* (Plate 21), which feeds on the same plant, instead has a pattern that conceals it when resting on a thick stem with wings furled, a most untypical posture for a geometrid. However, there are even fewer cases of moths matching vegetation that is *not* a larval food plant, if algae and lichens are discounted. The Drinker Moth seems to be the only British example. Too big and bulky to disguise itself as the grass blades on which its caterpillar fed, it mimics instead a bunch of yellowing leaves (Plate 73). Like other 'dead leaf' moths, its forewings have silvery central spots giving the appearance of holes.

In the classic image, a cryptically coloured moth sits in full view, protected only by its camouflage. In fact, only a minority of cryptic species rely solely on their appearance to safeguard them from predators: most also hide. However, unlike insects such as beetles, cockroaches and earwigs, where the delicate wings used for flight are protected by tough elytra, adult moths cannot bury themselves in the soil or other debris. Those that pupate underground must tunnel upwards after emergence, but once their wings are expanded these are too fragile for such activities. Nevertheless, moths conceal themselves very effectively in grass tussocks and other ground vegetation, in leaf litter and detritus, or in cracks and crevices. Here they are presumably less visible to birds, but more at risk from small mammals such as shrews, and carnivorous ground invertebrates (which might not hunt by sight in any case).

This behaviour is characteristic of many Noctuidae, especially the darts. Trying to find any of this subfamily at rest in the daytime would be a thankless task, except in urban areas where walls and pavements offer fewer

places of concealment. Many
other noctuids are occasionally
seen on tree trunks or fences, but
normally they tuck themselves
away out of view. Such moths
tend to be cryptically coloured in
a general-purpose fashion, not
resembling any particular object
or adapted to a specific resting
site. Dull, unobtrusive shades predominate – murky browns and
greys, beige, fawn and drab –
these homespun colours giving
rise to many of their English

Plate 22: Dull but unmistakable, the Mouse Moth *Amphipyra tragopoginis* resembles its namesake both in colour and habits, skulking in dark crevices. Wooden sheds are often used for aestivating. (Banffshire).

names: brindles and rustics, quakers and drabs (Plate 80). Markings likewise
tend to be unremarkable, consisting merely of lighter and darker mottling
and variegation. As a result, many of the species look confusingly similar.

In addition, noctuids that rest concealed often have a characteristic com-
pact posture. They fold up like a penknife, with one forewing overlapping
the other, sometimes almost completely. Presumably this enables them to slip
more easily into narrow spaces, and helps to reduce abrasion and damage.
Their wing scales are often glossy, thoracic crests are reduced or lacking, and
the moth is slightly flattened dorso-ventrally. The Chestnut and the Mouse
Moth *Amphipyra tragopoginis* (Plate 22) show these features well: to me
they have a vaguely cockroach-like appearance.

Disruptive coloration

This is a form of cryptic coloration based on optical illusion. Its effect is to
break up the familiar shape or outline of the object into unrecognisable
components. The disruptive effect is achieved by juxtaposing contrasting
blocks or bands of solid colour, or with a strong continuous line that serves
to cut the object in two. Often counter shading is employed, to interfere with
the three-dimensional perspective as well. Most camouflage relies on disruptive elements in the pattern, among mammals the Tiger, Zebra and
Giraffe being prime examples, likewise modern army battle-dress.

It is hard to think of a cryptic moth that does not employ disruptive coloration to some extent. The Grey Carpet (Plate 129), resting on bare earth
resembling a whitish stone (GM Haggett, personal communication) is one
example; the wholly black industrial melanic forms of various species are
others. However, there are a few moths that rely entirely upon a disruptive
pattern for their camouflage, or so it appears to us. It would be difficult to
prove experimentally whether or not this interpretation is correct even in

Plate 23: The striking, clear-cut pattern of the Clouded Border *Lomaspilis marginata* is probably disruptive, breaking up the shape of the moth to confuse predators. However, it may well be unpalatable too. (Banffshire).

Plate 24: Blood-vein *Timandra comae*. In the natural resting position, the strong line through each wing joins up to 'cross out' the shape of the moth – an effect that is lost in set specimens. (Sussex).

the laboratory, let alone in the wild. Therefore the following examples are only tentative, as seen through human eyes rather than those of the moth's natural predators. The Clouded Border *Lomaspilis marginata* (Plate 23) is certainly a shape-destroying medley of liver-brown and white. It rests openly on the upper surfaces of leaves. Similarly black and white, resembling a paper doily on a dark oak table, is the day-flying Argent and Sable *Rheumaptera hastata*. Several of the carpets have a whitish ground colour with a dark basal patch and central band; these include the Silver-ground Carpet *Xanthorhoe montanata* and the Wood Carpet *Epirrhoe rivata*. They are easily disturbed in the daytime, and rest fairly openly and indiscriminately, usually on vegetation. Perhaps their camouflage is disruptive too, but there may be an aposematic element: such moths are probably among the less palatable, without being toxic.

The Blood-vein *Timandra comae* (Plate 24) is the standard British example of a moth whose shape is bisected and thereby destroyed by a strong single line, but there are various others, such as the Vestal *Rhodometra sacraria*. Many other species incorporate two such lines, for example the Oblique Striped *Phibalapteryx virgata*, the Light Emerald *Campaea margaritata* (Plate 108) and the Scarce Silver-lines (Plate 12). Then there are numerous instances where a disruptive line forms a lesser element in the general cryptic pattern, as in the Orange Sallow *Xanthia citrago* (Plate 76) and the Rosy Rustic *Hydraecia micacea*.

Warning coloration

Many diverse groups of animals employ chemical or physical defences to protect themselves from predators. They advertise this fact by their bright colours, especially red, orange, yellow, white and black, usually combined in a simple, sharply contrasting pattern that is easily recognised and remembered (Plate 25). It is believed that non-specialist predators, including birds, have an inborn aversion to warningly coloured potential prey that is strongly reinforced by the first unpleasant encounter. Perhaps humans also inherit this distrust. As a very small boy, I was genuinely fearful of my first Garden Tiger *Arctia caja*, yet determined to possess it. Carrying the prize home in my school cap took all my courage, but since I did not eat it, the experience was ultimately a pleasant one.

In general, aposematic moths are diurnal, or easily disturbed during the day. They are sufficiently tough and pliable to withstand an initial mauling before a predator discards them because of their noxious smell or taste. If these defences fail to work, the ensuing digestive upsets should dissuade the predator from repeating its mistake, and it is likely to avoid other moths of similar appearance. This has led to many distasteful species resembling each other so that they present a united front to predators, a phenomenon known as Mullerian mimicry. The numerous very similar burnet moths (Plates 70, 72, 124) are a case in point, as are the clearwings. Elsewhere in the world, both groups form a mimicry complex with moths in the family Ctenuchidae, which has no native representatives in Britain, and also with certain wasps.

Plate 25: Garden Tiger *Arctia caja*. A familiar but still dramatic example of a warningly coloured moth, especially when the scarlet hindwings are displayed. Note also their metallic blue spots, this being a far rarer colour in moths than in butterflies. (Banffshire).

Rothschild (1985) gives full details of the toxins that have been isolated and analysed from British moths, but so far only the more obvious species have been properly investigated. Often they contain not one but several complex and highly toxic substances, each presumably effective against a different predator. Some toxins are sequestered and stockpiled by the caterpillar directly from the chemical defences of its food plant, others are metabolised by the insect itself. For example, the burnets synthesise hydrogen cyanide whereas the tigers produce histamine; both groups sequester pyrazines. Without specialised laboratory facilities, this type of research is beyond the capacity of amateurs, but ample scope remains to observe and document the behaviour of the moths themselves, and any inter-reactions with their predators that are witnessed in the wild.

The tussock moths also adopt a mainly aposematic lifestyle as adults. Several, like the Brown-tail *Euproctis chrysorrhoea* and Yellow-tail *E. similis*, synthesise histamine. They confirm how pure white, almost unmarked wings can act as warning coloration, as in the Large White *Pieris brassicae* butterfly and the White Ermine. Zigzag markings are present in the Black Arches *Lymantria monacha* (Plate 123), and these could easily be regarded as disruptive. It is likely that they do indeed fulfil a dual role. Even the most aposematic species are at risk from specialist predators able to withstand their chemical defences, or very hungry ones prepared to disregard an unpleasant taste. Thus their safest strategy is not to be noticed at all, but if they are discovered, to be recognised immediately as unpalatable. This explains why the forewings of highly aposematic moths are so often a compromise, displaying a warning yet offering some degree of concealment when the moth is inactive. Tremewan (1985) points out that the burnets are similar in size and shape to the black seedpods of trefoils and vetches common in their natural habitat, and can be mistaken for them when resting on a dull day. The Yellow-tail, sitting on a hawthorn hedge, looks very like a fluffy white feather. South (1907–09) remarks on how inconspicuous the Cream-spot Tiger *Arctia villica* is when its yellow hindwings are concealed, and I myself invariably fail to notice the Wood Tiger *Parasemia plantaginis* (Plate 109) before it flies up from the heather at my feet.

Of the relatively few British moths that have gone over completely to the aposematic lifestyle,

Plate 26: Lunar Hornet Moth *Sesia bembeciformis*. This moth is almost a caricature of a large and dangerous wasp, more convincing at first glance than its model. However, the female's wings still bear a few relict scales, as well as cilia, which in the male have already been dislodged during flight. (Banffshire).

the Lunar Hornet stands out. From any angle, at rest or in flight, it looks and sounds like a large wasp. It has all the vespine mannerisms, pumping the abdomen menacingly as if about to sting and buzzing angrily if disturbed. Forced to approach a mating pair (Plate 26) closely for the sake of photography, I could not prevent my scalp prickling. Though my intellect reassured me that these were harmless moths, a deeper atavistic instinct screamed of danger. Curious, then, that the moth's otherwise perfect mimicry shows one obvious flaw: its transparent wings have retained their fringes (cilia), whereas these are absent in the Hymenoptera. Cilia are present in all the other clearwings, and in the bee hawk-moths *Hemaris*. This suggests that in moths they fulfil some aerodynamic function that is too important to discard.

It used to be thought that these wasp and bee mimics were an example of Batesian mimicry, where a harmless species fools predators by its superficial resemblance to a dangerous one. Rothschild (1985) shows that the true situation is more complicated. Although it lacks a sting, the Lunar Hornet Moth is probably distasteful to most birds, whereas the real Hornet is palatable once its sting is removed. Nevertheless, the pupae of the Lunar Hornet are heavily attacked in early spring by the Great Spotted Woodpecker, which employs considerable effort to chisel them out of the larval borings in the trunk of sallows. It might seem impossible for the pupa to be palatable but not the imago, but in a few species distasteful compounds are created from innocuous components while the moth develops within the pupal case. However, Rothschild also notes that birds specialising in aposematic prey are themselves usually brightly coloured and repugnant to carnivores, for example the bee-eaters, bulbuls and drongos. So the Great Spotted Woodpecker itself may well be aposematic, forming a Mullerian mimicry complex with the many other similar black, white and red woodpeckers, and immune to the Lunar Hornet's defences. This bird has also been recorded feeding large numbers of Lackey *Malacosoma neustria*, Black Arches and Gypsy Moth caterpillars to its nestlings (Cramp, 1985). These well-protected hairy species are refused by most other birds.

Hawk-moths in Britain are generally palatable, except for the Spurge Hawk-moth *Hyles euphorbiae*, a scarce migrant. Many are highly cryptic when at rest. They are also protected by their sheer size and relative strength, being too large as adults for most insectivorous birds to tackle. On the other hand, this does make them worthwhile prey for domestic cats, which often figure in reports from the general public of Convolvulus Hawk-moths *Agrius convolvuli*. Several species conceal warning patterns beneath their cryptic forewings, revealing them in a well-choreographed display when molested. The Poplar Hawk-moth and the Eyed Hawk-moth engage in slow but powerful wing-flapping as the aposematic hindwings are exhibited, but the Privet Hawk-moth *Sphinx ligustri* freezes in a rigid posture that gives the banded abdomen prominence (Plate 27). Presumably it is pure bluff: Rooks regularly

Plate 27: At rest (top left), the Privet Hawk-moth *Sphinx ligustri* is cryptic, trusting in its camouflage. If disturbed (top right), its forewings part to reveal the abdomen's red and black warning bars. If further molested (below), the moth adopts its full defensive display. The strongly barred hindwings are now exposed, making an already large moth appear larger. The vulnerable antennae are tucked out of harm's way, as are the forelegs, leaving no convenient 'handle' for a predator to seize. Wingtips and abdomen are pressed firmly against the bark. Most hawk-moths are palatable, so this whole routine is probably bluff. (Sussex).

ate those moths attracted to a lighted subway near Brighton, leaving only the wings behind. Whether the bee hawk-moths are genuine Batesian mimics, or unpalatable in their own right, is also uncertain. Perhaps it depends on the predator or the circumstances.

Apart from the obvious groups already mentioned, it is likely that various other moths have warning coloration that is recognised as such by their predators, but is not so apparent to the human observer. Conversely, others might appear aposematic to us when in fact their markings have a different purpose. Two large, robust moths with day-flying males illustrate this uncertainty. The Emperor Moth (Plate 52) has large eye-spots in the centre of its wings, but Rothschild does not include it in her account of aposematic Lepidoptera. I have never seen the Emperor engage in any obvious defensive display, nor does it 'play dead' like other warningly coloured species. It is certainly eaten by birds including the Merlin and Stonechat. Ford (1955) believed that the eye-spots protect the moth while the wings are being inflated after emergence, when puckering emphasises them, giving the moth a terrifying appearance at a vulnerable time. On several occasions I have set up a camera to capture this event on film, but did not take a single photograph, there being no stage when the eye-spots were unusually prominent. The apparent lack of a warning display in the Emperor is even more puzzling when it is compared with the Kentish Glory, a moth of similar size, build and habits but with a cryptic disruptive pattern instead of eye-spots (Plate 82). Surprisingly, the male in particular does behave like an aposematic species: when molested it lies inert and displays its tawny orange hindwings and abdomen to best advantage.

Elsewhere in the world, some members of the Noctuidae are warningly coloured, like the Hieroglyphic Moth *Diphthera festiva* from tropical America, only on the British list as an importation. Almost all our indigenous species are palatable and cryptic, the most obvious exception (before it became extinct here) being the Spotted Sulphur *Emmelia trabealis* with its ladybird-like size and pattern. Especially in the tropics, the Geometridae include numerous brilliantly coloured

Plate 28: Chimney Sweeper *Odezia atrata*. Black perhaps functions as a warning colour for this day-flying geometrid, but (as so often) evidence from field observations is lacking. (East Inverness-shire).

Plate 29: Rannoch Brindled Beauty *Lycia lapponaria* is not usually considered to be an aposematic species, but shows several suggestive features including an orange dorsal stripe. The hairy, flightless female – little more than a sporran full of eggs – sits around in flagrant view. Here one is seen on top of a fence post much used as a perch by Meadow Pipits, as the droppings prove. (East Inverness-shire).

aposematic moths, like those in the genera *Milionia* and *Callioratis*. Again, European representatives are predominantly cryptic, although the Magpie Moth *Abraxas grossulariata* is an outstanding example of a species that is protected by warning coloration plus chemical defences in all its stages (Plate 42). The Chimney Sweeper (Plate 28), so conspicuous in its leisurely daytime flight, might also be distasteful. I have never seen it attacked by birds. In the Southern States of the USA, the diurnal White-tipped Black *Melanchroia chephise* is remarkably similar in shape and colouring, although not closely related. However, its thorax is orange, giving a stronger warning.

Evolution is not complete: it is happening now and will continue to do so as long as Earth exists. It follows that the species we see today are not necessarily the finished article: some will be midway between major changes in lifestyle and appearance. I suspect that the Rannoch Brindled Beauty is one. Most members of its genus are cryptic, but this species is well on the way to becoming aposematic. Its body is already protected by copious hairs, especially in the almost wingless female (Plate 29). One that I placed on a busy bird table survived for 40 minutes before being carried away (but not necessarily eaten) by a Coal Tit. The male has quite a marked defensive display, opening its thinly scaled wings wide while 'playing dead'. In the far distant future, perhaps it will have evolved into a 'clearwing' bee mimic, with the narrow orange dorsal stripe that already runs head to tail having become a broader and more noticeable feature. Becoming wholly diurnal (I believe it is already partly so) would surely benefit such a northern upland species that flies in early spring.

Bird-dropping mimics

Resemblance to a bird-dropping is usually considered to be a form of crypsis. However, although the moths are disguising themselves as a natural object, it is a highly distasteful one, repugnant to all but a few specialist coprophagists able to cope with the uric acid content and other concentrated excretory products. The corrosive effect of bird-droppings on car bodywork and even glass is well known to householders. Therefore, I suggest that such moths should be regarded as employing a type of aposematic rather than cryptic coloration.

Birds have a cloaca, where the concentrated urine from the kidneys joins with faeces from the alimentary system. These liquid and solid elements, creamy white and dark brown/grey respectively, are clearly discernible in their droppings. Moths that mimic them incorporate both these colours in their pattern, often with additional blue-grey and ochreous brown shades to represent the slight merging that takes place. Fresh droppings are moist,

Plate 30: Lime-speck Pug *Eupithecia centaureata*. An obvious bird-dropping mimic, left alone by birds. (Norfolk).

and slimy with mucus. In the moths, silvery or bluish metallic scales give this effect, as in the Chinese Character *Cilix glaucata* and Clouded Magpie *Abraxas sylvata*. This faithful attention to detail distinguishes the true bird-dropping mimics from other dark brown and white moths, especially when their habits are taken into account. Thus the Purple Bar *Cosmorhoe ocellata* and Blue-bordered Carpet *Plemyra rubiginata* qualify on appearance and sit openly on the upper surface of leaves, whereas the Silver-ground Carpet is probably disruptively marked and usually rests in a less-exposed position.

Size is clearly a factor governing successful resemblance to a bird-dropping. Apparently, the moth should not be too large, since the strategy is commoner among micros. Species in several genera of tortricid moths use it. They are inelegantly known as the 'bird-craps'. Of these, *Epiblema cynosbatella* will be familiar to most gardeners by sight, if not by name, since it feeds on cultivated roses. The Clouded Magpie is exceptionally large for a bird-dropping mimic, but none the less effective. Those I used to find on Bracken growing beneath Wych Elms resembled the splashy excretions of pigeons and corvids, discharged from the canopy high overhead.

There is abundant evidence that bird-dropping mimics avoid predation by birds. Goater (1992) describes how the Chinese Character and Lime-speck Pug *Eupithecia centaureata* (Plate 30) rest unmolested around moth traps while insectivorous birds eagerly gather the harvest of other species. However, although we can observe how birds behave, we cannot ask their reasons. If we could, I suspect the reply would be: 'Of course we recognise them for moths, but they don't just *look* like a piece of you-know-what: they taste like it too!'

Whether these moths do indeed contain toxic compounds does not seem to have been investigated. Since almost every substance is harmful in excess, it is the levels of any such compounds that would be significant. The food plants of several bird-dropping mimics certainly have a high toxin content, offering those moths an opportunity to sequester them. Like the tiger moths, the Lime-speck Pug frequently uses ragworts. The Bordered Pug *E. succenturiata* uses Mugwort, a source of the potent absinthe. The food plant of the Scorched Carpet *Ligdia adustata* (Plate 63) is Spindle, in the same genus as the exotic *Euonymus japonicus* – plants that only a few specialists can eat, and a favourite of the highly distasteful Magpie Moth.

Flash coloration

This is mostly seen in diurnal moths, or in those that use flight to escape if disturbed in the daytime. Their hindwings are brightly coloured, but hidden beneath duller forewings when the moth is at rest, only becoming visible when it flies. Many are called 'Underwings' for this reason. In Britain, 30 or more species can be said to use flash coloration, depending on how strictly this is defined. Most of these species belong to the Noctuinae and Arctiinae. Among the Geometridae, only the Orange Underwing *Archiearis parthenias* and its sister species the Light Orange Underwing *A. notha* qualify. The Carnation Tortrix *Cacoecimorpha pronubana* is one of few instances among the micros.

There are various ways that moths can benefit from this tactic. The sudden flash of colour might cause a would-be predator to hesitate, especially since yellow, orange or red often denote a distasteful species. Many of the tiger moths (Plates 25, 109) have hindwings of one or other of these shades, perhaps keeping their most striking warning colours in reserve to be deployed only when urgently needed. The Red Underwing *Catocala nupta* and others in that genus have equally vivid hindwings hidden beneath forewings that are very cryptic against bark, together with a strongly banded underside (Plate 31). Rothschild (1985) regards them as aposematic (but not distasteful) species since the hindwings are a startle device, designed to nonplus a predator. When the moth settles again after a short flight, its bright elements are hidden, so it effectively vanishes. Among birds, the Hoopoe Lark and Stone Curlew apparently employ the same technique, but clear-cut examples are hard to find as conspicuous wing

Plate 31: Red Underwing *Catocala nupta*. On the upperside, cryptic forewings hide scarlet hindwings, but the strikingly banded underside is part of its flash coloration too, though less often illustrated. (Sussex).

markings could have evolved for display or communication with other flock members instead.

However, flash coloration in other moths seems unlikely to be aposematic, nor are the species that use it known to be protected by an unpleasant taste. The various 'yellow underwings' are good examples. There are six British species in the genus *Noctua*, two in *Anarta*, plus the Small Yellow Underwing *Panemeria tenebrata*. Here, the effect of flash coloration is to make the moth hard to follow in flight. Quite how this operates is unclear, but presumably some sort of optical illusion is involved. In all these species, the dull forewings are cryptic, quite different from the striking hindwings – vivid yellow or orange, edged with a black band. As the moth travels forwards with flickering wings, the rapidly alternating bright and dark colours perhaps cancel each other out, affecting the registration of the visual image on the retina, or the processing of it in the brain. Certainly I find it almost impossible to track moths such as the Small Dark Yellow Underwing and related species in flight, losing sight of them far more readily than similarly sized moths without flash coloration. However, as always, we should not assume that the vision of their natural predators corresponds with ours.

Other species have a less dramatic flash coloration. Several partially diurnal Noctuids in the subfamily Heliothinae have pale hindwings with a wide dark brown outer margin more or less sharply defined. The Antler Moth and the Silver Y *Autographa gamma* have a suggestion of this pattern too. Curiously, the Straw Underwing *Thalpophila matura* appears to have excellent flash coloration yet it rarely flies in the daytime. Perhaps it did in the past, or is developing that lifestyle: evolution being a continuous process.

Melanism

Melanin is the pigment responsible for many of the brown and black colours in Lepidoptera. However, certain moths have genetically controlled varieties with such a saturation of this pigment that their normal colours and markings are obscured or completely overwhelmed. These are known as melanic forms. They occur in a very wide range of species, but usually they are rare. However, after the Industrial Revolution, melanic forms of many moths became increasingly common in manufacturing areas, especially in the cities of the north of England. Pollution of the environment by smoke and other noxious discharges was soon suggested as the probable cause, but there was dispute as to the way it operated. Some biologists held that genetic mutations had arisen from caterpillars eating foliage contaminated with metal salts, others believed that predation was involved.

The Peppered Moth and its almost totally black form *carbonaria* (Plate 32) epitomise the dramatic spread of an industrial melanic form through the population, almost to the complete exclusion of the normal type in some

Plate 32: As can be seen, the simple pepper-and-salt camouflage of the Peppered Moth *Biston betularia* in its typical form (above) is surprisingly effective against bark covered with crustose lichens: foliose lichens are not essential for its concealment. (Banffshire). Its famous industrial melanic form *carbonaria* (below) is rather conspicuous against clean bark, probably explaining its recent decline. (Sussex).

regions. First recorded in Manchester in 1848, f. *carbonaria* comprised 98 per cent of the population of some northern cities within 60 years (Kettlewell, 1973). This was evolution in action, and at a much faster pace than had previously been thought possible. In a classic series of experiments, Kettlewell showed that differential predation by birds was the key factor. In industrial areas the typical whitish form of the moth was poorly camouflaged against the soot-blackened bark of trees, so it was more quickly found and eaten, whereas the melanic form was cryptic. In unpolluted woodland, the situation was reversed: here the whitish form was well-concealed against the lichened bark, but the melanic one stood out and suffered heavier predation.

While this basic thesis is still generally accepted by evolutionary biologists, various flaws in Kettlewell's experiments have since been pointed out. These are discussed by Majerus (1998); in fact, Kettlewell was aware of most of them himself. His Peppered Moths were placed by hand on selected tree trunks, whereas their natural choice of resting site is much higher, often on the shaded underside of large boughs and branches in the canopy. They were also presented at an abnormally high density, several to each trunk, potentially encouraging birds to specialise in them. Assessments of whether the moths were cryptic against different backgrounds were made using human vision, but birds can see more ultraviolet light, which certain crustose lichens reflect strongly, potentially altering the degree of contrast. In calculating mathematical models to quantify the relative advantages of the two forms, Kettlewell did not allow for migration or dispersal, both by adults and by wind-borne tiny caterpillars on their silk threads. Mated pairs stay together for about 20 hours, and a predator spotting one of the moths is almost certain to see its mate too. In mixed pairs this would influence the mortality rate of the more cryptic form through no fault of its own. Finally, subsequent workers were not always able to reproduce Kettlewell's results. However, this is not surprising, as – appallingly – they tended to use dead moths glued to trees! At least Kettlewell's moths were able to orientate themselves and shuffle down into a comfortable resting position, as he describes them doing.

Perhaps this proves only that ecology is not an exact science. Almost any complex field experiment will have flaws: the very fact of its being a contrived set of circumstances is one. Then there are too many incalculables. The predation rate of melanic Peppered Moths, or of any other species, will vary even in the same part of the country. It will depend on the age and structure of the wood, the density of its bird population and whether pairs are feeding hungry young, the availability of alternative prey, and the weather on that particular day. In the circumstances, Kettlewell's methods and results were remarkably robust and convincing.

Those who did not experience it may not appreciate the severity of atmospheric pollution in industrial and urban areas before measures were taken to reduce it. For the first 18 years of my life I lived in Salford, part of the Manchester conurbation. Only after moving to Sussex in 1964 did I realise that black is not the natural colour of nasal mucus. The lumps of dried snot that we small boys used to hook out of our nostrils were known locally as 'crows' for that reason. Everything was black, coated in a layer of soot and grime: buildings, fences, tree trunks, and even the older leaves of evergreen shrubs such as privet. Most of the smoke came from domestic chimneys. Economic circumstances were hard, especially in the decade after the Second World War. Many households could afford only low-grade coal with a high sulphur content, and fires were burning all day long because few mothers went out to work. Power stations and railway marshalling yards contributed more smoke, while heavy industry discharged a cocktail of far more sinister and noxious substances into the atmosphere. For good reasons, the stretch by the canal where we used to play was known as 'Forty Stinks'. Certainly there were no lichens. After a trip to the countryside, I brought home a branch festooned with lichens and kept it in the garden, but they all died within a few months.

In such a setting, it was hardly surprising that many of the moths were black. In Salford, about 30 macro-moths commonly had melanic forms. Although Kettlewell does not mention any industrial melanics among the microlepidoptera, *Diurnea fagella* was often coal-black too. It seemed patently obvious that the dark forms had been favoured because they were less conspicuous to birds. Whilst the obvious explanation is not always the correct one, in this case I remain convinced that it was. All the melanics were species that habitually rested openly on walls, fences and tree trunks where they would be vulnerable to predation if poorly camouflaged. None of the moths that normally sat on foliage was melanic, nor was any aposematic species. This was particularly noticeable in the subfamily Ennominae, to which the Peppered Moth belongs. The Scalloped Hazel *Odontopera bidentata*, Pale Brindled Beauty, Willow Beauty, Small Engrailed *Ectropis crepuscularia* and species with similar resting habits were commonly melanic. However, those hardly ever found on tree trunks, like the Magpie, Brimstone Moth *Opisthograptis luteolata*, Swallow-tailed *Ourapteryx sambucaria*, Common White Wave *Cabera pusaria* and Light Emerald, showed not the slightest tendency to produce dark forms. If some factor other than predation were responsible for melanism, for instance a greater physiological hardiness in adverse conditions, then the resting site of the species would be irrelevant.

Since Kettlewell's time there have been fascinating developments. Following the Clean Air Acts from 1964 onwards, and the decline of traditional heavy industries, air pollution has been greatly reduced. Returning to Salford in the winter of 1997 after a long absence, I was amazed at how much cleaner everything was, including the atmosphere. Previously, the permanent smoke haze had always limited visibility; now, distant panoramas formed an unfamiliar backdrop to parts of the town I had known well. The Peppered Moth in particular has responded to these changes with remarkable speed. Form *carbonaria* is now in retreat. While this might appear to vindicate the predation hypothesis, since trees are no longer blackened by soot, authors such as Coyne (1998) instead consider it casts doubt, because f. *carbonaria* began to decline before foliose lichens reappeared. However, the typical form of the Peppered Moth is extremely well camouflaged against crustose lichens, which return much sooner than foliose lichens, or even against clean bark. Grant *et al.* (1998) show that there was a parallel rise and fall of melanism in the North American subspecies of the Peppered Moth in Michigan and neighbouring states, likewise correlated with levels of air pollution as measured by its sulphur dioxide content.

Melanism in moths also occurs in relatively unpolluted rural areas. Kettlewell (1973) attempted to categorise and explain this as well, but much less convincingly. His definition of what qualifies as a melanic form was far too broad. He included the strikingly disruptive black and white form of the Silver Cloud *Egira conspicillaris*, form *grisea* of the Pine Beauty (which is actually much *lighter* than the type), and the Scottish subspecies *myricae* of the Sweet Gale Moth *Acronicta euphorbiae* on the grounds that other European races were paler! Few other observers would agree with him – if a species has two or more forms it is almost inevitable that one will be darker, but that does not necessarily make it melanic in the generally understood sense.

Having devalued the concept in this way, Kettlewell understandably had difficulty analysing it. He postulated various kinds of non-industrial melanism, including Northern Melanism, Western Coastline Melanism, and Relict Conifer Melanism, illustrating these with selected examples but ignoring any species that did not fit. Thus the Tawny Shears *Hadena perplexa*, almost white in south-east Kent but blackish in Ireland, represented Western Coastline Melanism, but the Grey Arches which varies in the opposite direction was not mentioned. Likewise, moths such as the Dark Marbled Carpet *Chloroclysta citrata* are much whiter in northern Scotland than elsewhere in Britain. The Grey Pine Carpet, Pine Carpet and Tawny-barred Angle *Macaria liturata* are never melanic in the remnants of the Ancient Caledonian Forest that survive on Deeside and Speyside, but they are in southern English pine plantations. Moths do not always fit into neat categories to order.

Variation: the advantages of being different

To reduce direct competition, all living organisms try to monopolise their own separate niche. The structural and behavioural adaptations required will inevitably tend to promote differences in appearance. A large number of moths are associated with birch, and rest upon that tree. Yet because they choose different locations, they embody a great variety of shapes and colouring, appropriate to their habits. Of those that sit on the trunk, some select gnarled old bark, others the smooth silver areas, and are camouflaged accordingly. Some wrap their wings around smaller branches, or thin twigs. A few are green to match the foliage, others brown or orange to suggest dead leaves. However, species that do share a similar lifestyle and habitat are exposed to the same selective pressures. If they respond in the same way, they may come to resemble each other, at least superficially. This is known as convergence.

For aposematic species, convergence has clear advantages, since each can gain vicarious protection from being mistaken for another, preferably a more distasteful one. 'Deliberate' mimicry is therefore a feature of warningly coloured moths, as has already been noted. For palatable species that rely on camouflage, convergence may instead be dangerous. Predators hunt using a 'searching image', which has to be learnt. If cryptic species resemble each other too closely, a predator's familiarity with one will put the others at risk. As a result, natural selection may be pulling in opposite directions, some factors encouraging convergence and others favouring difference as a protection in its own right. Were it not for this, I suspect that many more of our moths would be almost indistinguishable in appearance from their relatives, and certainly less diverse.

The wainscots show convergence in action. These assorted genera of Noctuids have independently evolved straw-coloured, striated forewings that conceal the adults when they rest amongst the grasses and sedges

Plate 33: The Reed Dagger *Simyra albovenosa* (left) is in a different subfamily (Acronictinae) from all other British 'wainscot moths', but has nevertheless converged in appearance with various other reed bed and grassland species such as the Large Wainscot *Rhizedra lutosa* (subfamily Amphipyrinae) (right). Note how raised veins match the texture as well as the colour of the background reeds and rushes. (Norfolk; Banffshire).

that are the larval food plants (Plate 33). Equally, not all Noctuids with that lifestyle have adopted the wainscot pattern. Those in the genera *Apamea*, *Oligia*, *Amphipoea* and *Celaena* are every bit as much moths of grassland and marshland, but have a more general-purpose, mottled or banded cryptic coloration instead. Presumably they benefit from their dissimilarity.

Differences between species are only to be expected, but in some moths the sexes differ considerably too. This is known as sexual dimorphism. It may involve wing coloration, build and size, with the female usually being the larger. Male and female can be so unalike that early authors were fooled, and described each as a separate species, adding to the confusion of scientific names. Whereas in butterflies, sexual dimorphism is usually connected with display (as in birds), in moths it is associated with major behavioural differences between the sexes. It is common where one sex flies in the daytime and the other at night, as in the Oak Eggar and the Clouded Buff *Diacrisia sannio*. More extreme examples involve moths with flightless females, like the Vapourer and the Winter Moth (Plate 79). Sometimes the purpose of sexual dimorphism is less obvious, as in the Broad-bordered Yellow Underwing *Noctua fimbriata* (Plate 34). However, females of this

species aestivate, and perhaps the males do not: all those I have seen in late summer have been female.

Other variation affects both sexes equally. Sometimes it is so slight as to be barely noticeable, in other cases so extreme that it is hard to believe that individuals belong together. Nearly all variation is genetically determined, sometimes influenced by environmental factors such as cold or heat. Where several genes are acting in concert this is known as multifactoral variation, and tends to be continuous, different forms grading into each other. Where variation is controlled by a single gene, the forms produced can be strikingly disjunctive, not connected to the typical one by intermediates. Often, such forms are rare, in which case they are known as aberrations. However, if they form a significant part of the population (there seems to be no agreed minimum percentage) the species is said to be dimorphic, or polymorphic if more than one additional form is involved.

Ford (1945 and 1955) analyses variation in the Lepidoptera in detail, giving instances where the gene causing it is known to be dominant, recessive, sex linked, sex controlled, or occurs as multiple allelomorphs. He expressed the hope that all varieties of butterflies and moths will be investigated to determine their genetic characteristics, and urged amateur breeders to participate in this work. Largely, this has not happened. In any case, being able to identify the gene responsible for a particular variety does not necessarily take us very far forward. Ecology, rather than genetics, is now seen as the

Plate 34: Broad-bordered Yellow Underwing *Noctua fimbriata*. Although both sexes are variable, males are consistently darker than females (particularly so in Banffshire, as illustrated). It is unclear why a wholly nocturnal species that rests concealed should show such sexual dimorphism.

most important field of study. Such a change of emphasis is welcome in that discovering what a species requires and how it behaves has greater practical applications, particularly for conservation.

As a result, amassing collections of varieties, and naming each one if it differs ever so slightly from the next, is no longer in vogue. Admittedly, in the past the practice has been taken to excess. Frequently, several authors all named the same form, leading to a confusing array of synonyms, or faded specimens were claimed as new varieties. And certain species are almost infinitely variable: Manley (1973) described over a hundred forms of *Acleris cristana*, and almost as many have been listed for the related *A. hastiana*. The modern reaction against such pedantry is hardly surprising.

Arguably, the pendulum has swung too far the other way. Variation is important. Without it, natural selection cannot operate. Life would not have evolved beyond single, self-replicating molecules: there would be no moths and no humans to study them. If some moths are exceptionally variable, they must gain an advantage by this, otherwise the fractionally less-successful forms would have been eliminated. Variation, or lack of it, is a vital element in the make-up of a moth, and should be recorded. What else is a species but the product of its DNA, expressed through the action of its genes?

Too often for coincidence, our most abundant and ubiquitous moths are also the most variable, whereas localised, low-density species are usually very constant. Those with the greatest variety of forms include the Common Marbled Carpet *Chloroclysta truncata*, July Highflyer *Hydriomena furcata*, Heart and Dart *Agrotis exclamationis*, Large Yellow Underwing, Ingrailed Clay *Diarsia mendica*, Clouded Drab, and Common Rustic *Mesapamea secalis*. Others, although less widespread, are the most numerous moth in their particular habitat, like the Grey Pine Carpet and Coast Dart *Euxoa cursoria*. In contrast, almost every moth regarded as a Priority Species under the UK Biodiversity Action Plan (Bourn *et al.*, 1999), because it is local and scarce, shows hardly any variation, at least in Britain.

Inevitably, there are a few exceptions. The Winter Moth, Flame Shoulder *Ochropleura plecta* and Smoky Wainscot *Mythimna impura* abound almost everywhere, but vary little. The Oak Nycteoline has over 40 named forms, but is local and not particularly numerous, as are the two *Acleris* species mentioned above which it superficially resembles in appearance and habits: all these moths hibernate as adults.

One explanation for the tendency of abundant species to be variable is because of the very size of their populations. Genes that occur at a low frequency can survive, whereas in small populations they risk being eliminated by random genetic drift: if a gene was present in only 1 per cent of individuals, and the colony dropped to a few dozen adults, the odds are that the gene would be lost. Thus the gene pool shrinks, and inbreeding becomes more likely with all its deleterious effects. Visible variation is therefore a healthy sign, being the outward indication of genetic biodiversity, described by Ennos and Easton (1997) as the evolutionary potential to adapt successfully to future environmental change. Perhaps, with industrial melanism, we have already witnessed this. It is impossible to say whether trunk-resting moths without a melanic form available in their gene pool would have died out in the worst smoke-polluted areas, but virtually every species that survived did so as a melanic form.

Another possibility is that species whose numbers are consistently high enough to attract the attention of predators benefit from having a variety of different forms, often including plain and banded ones, as pointed out by Kettlewell (1973). This forces their enemies to learn many searching images rather than one, so it is harder for them to 'get their eye in'. I have great difficulty locating the Dark Marbled Carpet (Plate 35) and July Highflyer (Plate 36) on tree trunks even knowing roughly where the moth

Plate 35: Dark Marbled Carpet *Chloroclysta citrata*. Abundant species are often very variable. Four examples from the almost infinite variety of forms found at Ordiquhill in Banffshire are illustrated. The moth rests on tree trunks during August and into September, and perhaps its diversity prevents birds learning a single 'hunting image' to help them exploit such a good potential source of food.

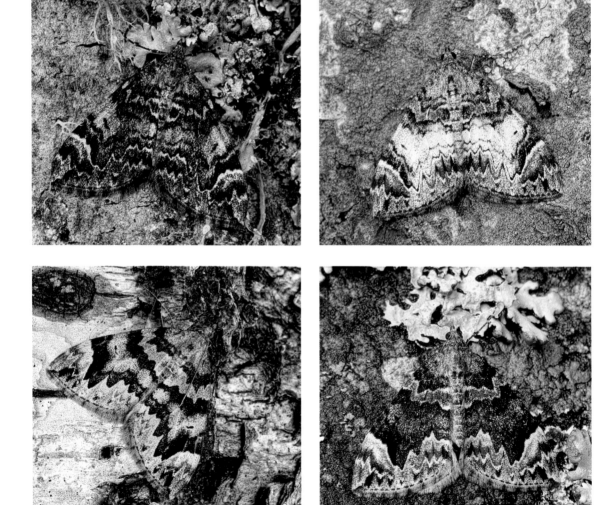

Plate 36: July Highflyer *Hydriomena furcata*. Four examples from the aforementioned Banffshire site are shown. Again, its plethora of forms may baffle searching predators. It flies at the same time of year as the Dark Marbled Carpet and is equally abundant, also resting on tree trunks. Significantly, the colours and patterns of the two species never overlap: if a predator learnt to recognise one, the other's camouflage would not be compromised.

has settled, so diverse are the colours and patterns of their multitude of forms. Whether each form chooses an appropriate background is less certain. However, Majerus (1998) provides data showing that morph-specific habitat preferences operate over distances of only a few metres for moths such as the Riband Wave *Idaea aversata*, the banded form being proportionately more numerous in shaded, closed canopy woodland and the plain form in more open areas.

The early stages

The population of every moth is at its most numerous in the egg stage, but eggs are tiny. It is at its lowest in the imago, but this makes a size-able meal. Somewhere in between, the population reaches its greatest biomass – the combined weight of all its surviving individuals. As we do not know detailed mortality rates for most (if any) species, it is impossible to pinpoint this moment exactly, but I suspect that it usually arrives when the caterpillars are between three-quarters and fully grown. (However, some will be already doomed because they harbour parasitoids.) This is therefore the time when moths offer predators the greatest potential source of food, and also the most convenient source. Caterpillars cannot fly away, or run very fast. Except for the hairy ones, the proportion of indigestible chitin (mainly the head capsule and thoracic legs) in a caterpillar is lower than in any other stage. Consequently, nearly all the protective features employed by adult moths to avoid being eaten are seen in caterpillars too. Some are also found in the egg and pupa, where these are exposed to view.

The egg

Eggs are often tucked away in crevices or hidden in vegetation. It is likely that invertebrates, rather than birds, are their main predators. Eggs that are in view for a long time are often cryptically coloured. Those of the Merveille du Jour *Dichonia aprilina* and many other autumn-flying moths overwinter, and are very inconspicuous on the bark and twigs where they are laid. The burnet moths lay bright yellow, aposematic batches. No doubt other protective features of eggs escape our notice because of their small size.

The caterpillar

Caterpillars are handy packets of protein, especially for birds. They are easily transported by the beakful to hungry nestlings. As a result, they are a major item of prey for almost every insectivorous bird – even Swallows take those dangling from threads – and often they are numerous enough to be deliberately targeted. Fortunately they can employ a wider range of strategies than adult moths. Having jaws, they can tunnel inside their food plant. Able to produce silk, they can construct habitations for themselves. Those of the microlepidoptera generally inhabit leaf-mines, cases, borings or spinnings. Caterpillars of some macro-moths bore into stems or spin leaves together; others feed only at night. Thus most depend partly on concealment to reduce predation, at least from birds. Relatively few feed or rest in full view during the daytime, but those that do provide many striking examples of camouflage and aposematic coloration.

For cryptic species, eliminating the familiar sausage-like appearance of the average caterpillar is clearly a high priority. In many noctuids, this is achieved without any

Plate 37: The spiral posture adopted by these Brindled Pug *Eupithecia abbreviata* caterpillars helps them to resemble shrivelled bracts of the oaks on which they feed. (Aberdeenshire).

Plate 38: Like many geometers, the caterpillar of the Smoky Wave *Scopula ternata* is almost impossibly long and thin, easily overlooked as a stiff dry stalk of its food plant. (Aberdeenshire).

Plate 39: Dorsal humps, posture and markings all combine to destroy the familiar 'sausage shape' in this Pebble Prominent *Notodonta ziczac* caterpillar. Linnaeus' specific name is particularly apt. (Banffshire).

structural modifications by employing disruptive markings, usually including a broad subspiracular stripe and often dorsal and subdorsal lines, dashes or chevrons too, paler or darker than the ground colour (Plates 95, 102). In other families, the actual shape is considerably modified – unlike the imago a caterpillar does not need to be aerodynamically efficient. Resting posture – angular, coiled or sometimes spiral – also helps the deception (Plate 37). In many Geometrids, especially waves in the genus *Scopula* (Plate 38), the caterpillar is almost impossibly long and thin, likewise in the Ennominae where the resemblance is to a gnarled twig (Plate 92). The prominents are particularly varied. Often their caterpillars have dorsal humps, and rest with their front and back ends raised. The resulting jagged outline explains the scientific name of the Pebble Prominent *Notodonta ziczac* (Plate 39).

Caterpillars have the opportunity to resemble their food plant closely. This can be achieved in several ways, even by polyphagous ones. Some are translucent when young, so that the contents of the gut show through, making them the same colour as their food. Majerus (1998) illustrates this in the Angle Shades. Others, like a chameleon, have pigment cells that respond to some extent to their background. I once found a November Moth *Epirrita dilutata* caterpillar that was maroon like the young Wych Elm leaves on which it was feeding, but as the expanding leaves gradually became entirely green, so did the caterpillar. Those confined to a single food plant can of course tailor their camouflage to it (Plates 88, 94, 105, 116, 118).

Variation occurs in the larval stage too, and some caterpillars are polymorphic. Many species have a brown form and a green form, with the markings otherwise similar. Caterpillars of the December Moth *Poecilocampa populi* and Scalloped Hazel resemble brown twigs, but have a 'lichened' alternative in some areas, strongly variegated with white. Melanics are also known. In Salford (during the 1960s), Green-brindled Crescent *Allophyes oxyacanthae* caterpillars were black, as illustrated in Ford (1955), like the sooty bark where they rested when not feeding. Sexual dimorphism is apparently absent in caterpillars, although those that will produce female moths can often be recognised by their size when that sex is considerably larger than the male.

Aposematic colouring is relatively more frequent in caterpillars than in adult moths. Universal in families like the tussocks and tiger moths, it is also employed by many species that will give rise to highly cryptic adults, including all members of the large *Acronicta* genus of noctuids (Plates 40, 41), the Buff-tip (Plate 115), and the Mullein Shark (Plate 90). However, the converse is unusual: aposematic adults hardly ever have cryptic caterpillars. That the bee hawk-moths do is strong evidence that they are primarily Batesian mimics.

Although those of burnets and the Magpie Moth are exceptions, many aposematic caterpillars are densely hairy. Besides being a physical defence, the hairs are frequently glandular and barbed, charged with histamine. Even to humans, they can cause intense and prolonged irritation. Often bright colours reinforce the warning message at close range while still providing disruptive camouflage when seen from a greater distance. Caterpillars of the Pale Eggar *Trichiura crataegi* and Dark Tussock (Plate 89) can be surprisingly hard to spot amongst heather.

Both Peach Blossom *Thyatira batis* and Alder Moth *Acronicta alni* caterpillars mimic bird-droppings until their final instar (Plate 41). They sit on the upper surface of a leaf in a similar 'question-mark' pose, an example of

Plate 40: Sweet Gale *Acronicta euphorbiae*. Red, yellow and black warning colours, plus tufts of barbed hairs, mark this out as a typical aposematic caterpillar, yet it gives rise to a highly cryptic imago. (Banffshire).

Plate 41: Peach Blossom *Thyatira batis* caterpillars: fourth instar (top left), final instar (bottom left). (Banffshire). Alder Moth *Acronicta alni* caterpillars: fourth instar (top right), final instar (bottom right). (Norfolk).

convergence in appearance and behaviour as they are not closely related. However, the white uric acid portion of the dropping is at different ends: maybe it does not matter. In their final instar, they are presumably too large for the deception to work. They then employ diametrically opposite solutions. The Peach Blossom becomes fully cryptic and hides low on its food plant, whereas the Alder Moth assumes extreme warning coloration and continues to rest exposed. How the latter's switch could evolve is rather puzzling, since any intermediate stage would surely fall between two stools: neither a realistic bird-dropping, nor effectively yellow and black.

The pupa

Most pupae are well hidden, so do not need protective coloration. The chitin of their shell is usually red-brown or blackish, presumably for incidental or structural reasons. The few that are exposed, held in place by a few strands of silk or an open mesh cocoon, are generally cryptic. Aposematic pupae, as in the Magpie Moth, are rare (Plate 42).

Plate 42: Magpie Moth *Abraxas grossulariata* caterpillar and pupa. Few British geometrids have aposematic caterpillars, let alone pupae. The pupal case itself is black, but transparent panels allow the body contents to show through, giving the appearance of yellow bands. (Aberdeenshire).

Conclusions

It may be no coincidence that regular predators on moths rarely specialise on resting adults: if not distasteful, they are too well hidden for this to be worthwhile. Dragonflies catch flying moths, and spiders net them. Bats catch moths at night using echo location rather than sight. Birds also tend to catch them in flight, when the moths' cryptic coloration is less effective or irrelevant. Nightjars hawk moths at dusk, while Little Owls profit from lekking male Ghost Moths. Common Gulls take many Autumnal Rustics *Eugnorisma glareosa* during the light Shetland nights (Kettlewell, 1973). Falcons such as the Hobby and Merlin catch large day-flying Oak Eggars and Emperors on heaths and moors respectively. Whilst numerous other birds eat adult moths opportunistically if they happen to notice them, perhaps only the Treecreeper habitually searches for those resting on tree trunks and branches.

Unlike caterpillars, it seems that adult moths are not particularly edible. Wings are normally removed, adding to handling time. And it is probable that many cryptic as well as recognisably aposematic species contain distasteful compounds: birds discard the abdomen of the Beautiful Golden Y. Perhaps this explains why bird predation seems to be highest during the breeding season. If adult birds themselves do not eat many moths, they certainly feed them to their young! Interestingly, newly hatched birds have far fewer taste buds than adults – in chickens, only a third as many – (O'Connor, 1984) so presumably they do not mind. Although it is not an entirely natural situation, the birds in my garden almost completely ignore moths outside the trap from March until late May. After this, most are quickly picked off at first light, and fed especially to fledglings. Interest wanes around mid-August, once the young are independent. Since the main culprits are the resident, experienced pairs of Blackbird and Song Thrush, known by their rings to be present continuously, the seasonal pattern of predation appears genuine.

Chapter 3

Numbers and distribution

In global terms, the number of species found in a given area is highest at the equator, and declines towards the poles, independent of habitat diversity. Thus we have the vast multitude of species (but often with small populations) in the tropical rain forests, compared with very few at the poles. However, those that do occur at high latitudes tend to be found in huge numbers, like the penguins of Antarctica or the Long-tailed Duck and Little Auk in the Arctic. Numbers of species and individuals in temperate zones fall somewhere in between these extremes.

The British list

Karsholt and Radowski (1996) provide a table showing the number of Lepidoptera (excluding adventives) found in each country in Europe. France tops the table with 4,755 species, closely followed by Italy and Spain. Totals diminish northwards; predictably, Iceland is at the bottom with only 92 species. Britain, with 2,386 species, perhaps has fewer than would be expected from its latitude and area. For instance, it is out-scored by Denmark with 2,423 species and narrowly by Latvia with 2,399, yet both countries do not extend as far south and are only a fraction of Britain's size. However, Britain is an island off the north-western edge of the continental land mass, and the sea acts as a barrier to the more gradual forms of colonisation or range extension. If we ever did become 'part of Europe' we would have more moths!

The pattern within Britain

In Britain, the number of species is highest in the south and east, as Figure 1 and Appendix 1 show. Adequate information is only available for macros,

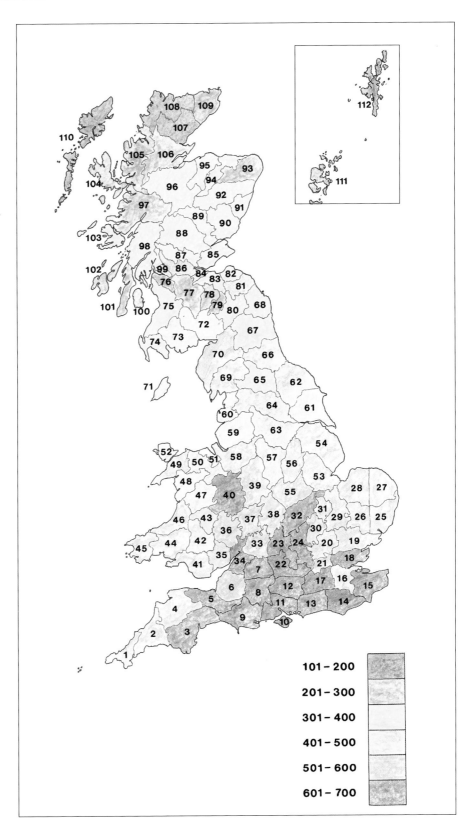

Figure 1: The Watsonian Vice-Counties of Great Britain, with the number of species of macro-moth recorded from each. (See Appendix 1 for key to vice-county names, actual totals or estimates, sources and methods.) Redrawn from *The Moths and Butterflies of Great Britain and Ireland* vol. 1 (1983) by kind permission of Basil Harley.

101 – 200

201 – 300

301 – 400

401 – 500

501 – 600

601 – 700

but of the Watsonian vice-counties (fixed biological recording areas of rough-ly equal size (Dandy, 1969)) Dorset is the richest with 687 species. The geol-ogy of Dorset is very varied: there is both chalk and limestone, as well as sandy heaths. It also receives a good share of rare immigrants from across the English Channel, especially at Portland. Next comes South Hampshire with 678 macro-moths, partly thanks to the New Forest, even though this is no longer quite the Mecca for lepidopterists that it was in Victorian times. Then, several hundred collectors descended on the area annually, bagging the best sugaring rides by pinning their visiting cards on trees to the annoyance of the locals (Oates, 1996). Those days are gone, but it still has a wide range of scarce species.

At the other end of mainland Britain, 900 km north, Caithness has only 248 macro-moths. This does not mean that it has 'lost' 439 of the species recorded in Dorset: the actual number missing is actually higher, but in com-pensation Caithness has some northern and upland moths that Dorset lacks. East and West Sutherland's dismal totals reflect under-recording: doubtless a keen observer could add many new species to either list. The Northern Isles, on the other hand, have always attracted interest because of their dis-tinctive races of many common moths. As with its birds, Shetland's total is boosted by migrants from the Baltic or Fennoscandia when the wind is east-erly. Recent work by the enthusiastic Shetland Entomological Group has led to some very surprising moths being recorded, as well as the Swallowtail but-terfly *Papilio machaon* – not once, but twice! (Pennington, 1996).

The underlying trend of fewer species northwards is of course modified by other factors. Counties with a coastline generally rank higher than inland ones. First, they receive more migrants; secondly, some moths are largely confined to coastal sites such as cliffs, shingle, saltmarsh or sand dunes. Then the quality and variety of habitats in a county obviously matters, with ancient mixed deciduous woodland particularly important. This may explain why Shropshire and Nottinghamshire (with Sherwood Forest) stand out as particularly rich. Geological diversity helps too, especially if there is chalk or limestone. Both are good for moths, and quite a few species are restrict-ed to such areas even though their food plants are not. Partly because of this, Berkshire, Oxfordshire and Northamptonshire have exceptionally high totals for inland counties.

Further north, the Lake District has always been favoured, again because of its varied geology and topography, and its relatively unspoiled habitats. Botanically rich Durham is good for moths too. In Scotland, a paucity of data hinders comparisons. Arran, Rum, and even the Perthshire vice-coun-ties, all renowned hunting grounds for collectors, lack collated totals. Estimates suggest that Berwickshire and Stirlingshire are above average. South Aberdeenshire, which includes many wonderful habitats on Deeside, predictably has a good score, and Moray's list is exceptional because it takes in Speyside as far as Aviemore, as well as favoured coastal areas at Findhorn and Culbin.

Conversely, there are counties with smaller than expected scores. West Cornwall and East Cornwall are our southernmost vice-counties, and both are renowned for rare migrants, but many resident moths (and birds) peter out before reaching here. Perhaps the climate is just too maritime: too warm, windy and wet for the more 'continental' species. East Anglia is synonymous with intensive arable agriculture – prairie farming – to the detriment of its wildlife. It still has many rare specialities among its moths, particularly on what is left of the Breck and Broads, but the overall species lists of both East and West Norfolk are relatively disappointing considering their past reputation.

Distribution and ecology

At the level of the individual species, the distribution of moths is endlessly fascinating. Except for a few special studies on species of high conservation priority, we have little hard information on the subtler requirements of most moths beyond the obvious presence of their food plant. It is intriguing, though, to look at distribution maps such as those in Heath and Emmet (1979), and speculate as to the reasons for a moth's presence or absence in certain areas. Confining attention just to the subfamily Noctuinae (as defined in that volume) the dot maps bring both enlightenment and puzzlement. There are the obvious coastal species like the Archer's Dart *Agrotis vestigialis* and the Portland Moth *Actebia praecox*; a dependence on food plants growing in sand or shingle is neatly confirmed by their few inland colonies where the terrain is suitable. For the Light Feathered Rustic *Agrotis cinerea*, geology is the key: the dots coincide with chalk and limestone areas in the southern half of Britain. The Shuttle-shaped Dart *A. puta*, so abundant and at least double-brooded in southern England, reaches a surprisingly well-defined limit, being virtually absent north of the Humber. Perhaps it cannot adopt the univoltine life cycle that would enable it to extend further north. Conversely, the Northern Dart, a classic arctic-alpine species with a two-year life cycle, is confined to mountaintops in Scotland and northern England. Other moths like the Flame Shoulder and Large Yellow Underwing are found everywhere: it is the absence of observers rather than the moth that is responsible for the higher proportion of blank squares in Scotland.

Those were the easy ones, where we can surmise from its distribution what each species requires with moderate confidence. Others are more problematical. The Dotted Rustic *Rhyacia simulans* was, before its recent spread, found inland on calcareous soils in central southern England, and also sporadically on rocky coasts, moorland, and mountain scree as far north as Orkney. More dissimilar habitats are hard to imagine. The Plain Clay

Eugnorisma depuncta (Plate 43) has a strictly eastern distribution in Scotland and northern England, but a western one further south. Most baffling of all is Ashworth's Rustic *Xestia ashworthii* (Plate 126), confined to a discrete area of north Wales, where it is common. It does not seem to be restricted much by geology, and certainly not by food plant, for its caterpillar eats a very wide range of low plants. Rearing it in captivity is far from difficult, and it requires no special conditions. Other members of this genus have very wide distributions in Britain, and it is a mystery why this one does not extend beyond the area where John Ashworth discovered it in 1853. But then, the same could be said for Weaver's Wave *Idaea contiguaria*, which has a very similar British distribution.

Plate 43: Plain Clay *Eugnorisma depuncta*. Although found from Sutherland to Devon, the restricted distribution in Britain of this polyphagous woodland species is hard to explain either by climate, habitat or food plant. Its European distribution is also patchy. (Banffshire).

Perhaps, in time, the factors governing the distribution of our moths will be elucidated. I suspect that the climatic ones will prove the hardest to uncover. It is unlikely that these will be as simple as average summer temperature or total annual rainfall. The crucial effect might occur at any stage in the life cycle, each stage could have different requirements, or compensation might be possible – a cool spring being acceptable so long as autumn is warm. To invent a hypothetical example, a species overwintering as a caterpillar might require the temperature to stay above 7°C for most of the night on at least 20 days in October and in April, in 8 years out of 10. Finding that out would be almost impossible.

In any case, it is artificial to try to separate climatic and habitat requirements. The length of the growing period, the amount and seasonal distribution of rainfall, and average summer and winter temperatures obviously affect the type and lushness of the vegetation. Most species are able to trade one off against the other to some extent: where the climate suits them they are less fussy about habitat; where the climate is not ideal they can still survive if the habitat is perfect. Thus many moths that are ubiquitous common or garden species in the south of England become increasingly choosy further north. If they extend into Scotland they become indicators of the very best woodland or other prime habitat. Once it reaches Inverness-shire, the Poplar Grey *Acronicta megacephala* has long ceased to be the familiar urban species that it is over most of England and Wales, where it thrives on any sort of poplar, native or exotic. Instead, it is confined to natural Aspen woodland, as in the Spey valley, and in captivity the caterpillars refuse other types of poplar. The Old Lady, another town and garden species over much of Britain, is reduced at the northern limit of its range to its core habitat of sheltered lower river valleys, such as those of the Deveron and the Spey. The same principle can be seen in reverse. In north-east Scotland, the Brindled Ochre is usually the most numerous autumn noctuid, and its caterpillars can be found in the rootstock of almost any Hogweed growing on roadside verges or in abandoned lawns, whatever the terrain. Further south, it becomes increasingly associated with rocky ground, until on the south coast of England it is restricted to coastal cliffs.

Population size and density

Inextricably linked with a moth's distribution is its population size and density. There is a tendency for the most ubiquitous species to be the most numerically abundant as well, like the Large Yellow Underwing and the Dark Arches *Apamea monoglypha*. It is hard to find examples of species that occur almost everywhere in Britain, yet nowhere are seen in large numbers. The Herald (Plate 44) seems to be one, as pointed out by Young (1997); I have never seen more than three in one night in Sussex, Lancashire or Banffshire, though larger numbers sometimes accumulate at hibernation sites.

Plate 44: The Herald *Scoliopteryx libatrix* hibernates as an adult, thus it announces the arrival both of autumn and spring. Found throughout Britain, its low density makes a sighting always memorable, especially when its rich colours glow in the torch beam as it feeds on ripe blackberries or the sugar patch. (Banffshire).

Also puzzling are pairs or groups of species sharing the same habitat and food plant, and with similar lifestyles, but showing consistent differences in relative abundance. The Dark Arches and the Light Arches *Apamea lithoxylaea* are closely related grassland moths of similar size and habits. Both are found throughout Britain, yet in every part of the country the latter is by far the less numerous. Pratt (1981) gives data showing that it was outnumbered 31 to 1 by the Dark Arches over a 10-year period at Peacehaven in Sussex, even though Heath and Emmet (1983) describe the Light Arches' most favoured habitat as dry grassland in the south and east of England. On acid moorland in Banffshire, my equivalent figures are 60 to 1. In Scotland, the Tawny-barred Angle and the Bordered White are found in virtually every pinewood or plantation. They are in the same subfamily, are of comparable size, share the same parasitoids, and have a very similar life cycle. However, the Bordered White consistently outnumbers the other species by 50 to 1, even though populations of both show cyclical fluctuations (Barbour, personal communication). These ratios of abundance also occur between groups of species. In Ash woodland in Sussex, three specialists on that tree always maintained the same relationship from year to year. Commonest was the Centre-barred Sallow *Atethmia centrago*, second came the Dusky Thorn *Ennomos fuscantaria*, while the Tawny Pinion was always scarce. The latter is well known to be a low-density species.

There are at least two possible explanations for the differences in abundance described above. First, it might be wrong to assume that the habitat gave each species equal resources: perhaps only 1 in 40 grass tussocks is suitable for the Light Arches, whereas the Dark Arches can use them all. Similarly, the Tawny Pinion caterpillar could be restricted to leaves on the outer twigs of the third branch up, of an Ash tree at least 60 years old that is exposed to the full afternoon sun. Secondly, density dependent factors involving disease or predation might kick in much sooner and harder for some species than others. Perhaps a tree can only support a few Tawny Pinion caterpillars before birds begin to target them, whereas the threshold for Dusky Thorn caterpillars is much higher. One day these points might be resolved.

All the species mentioned so far are, where they occur, generally distributed – they are not confined to localised colonies. Obviously, if pine or Ash is the food plant, they will only be found where enough of those trees are present. If the food plant is uncommon and highly localised, so will be the moth: Touch-me-not and the Netted Carpet *Eustroma reticulatum* being one example. However, some moths having widespread food plants are nevertheless very local, often being abundant in small discrete areas but absent from apparently similar ground nearby. The burnet and forester moths are classic examples, as is the Scarlet Tiger. Others might include the Dark Bordered Beauty *Epione vespertaria* (Leverton *et al.*, 1997) and the Rosy

Plate 45: Many warningly coloured caterpillars are gregarious, like these of the Pale Eggar *Trichiura crataegi* (albeit photographed in captivity). Note the variation in size because of staggered hatching of the overwintering egg batch. In the north, late-hatching caterpillars can adopt a 2-year life cycle, this being one of very few British moths able to hibernate in different stages. (Banffshire).

Marbled *Elaphria venustula*. Butterflies, being day flying, make it easier to find examples such as many of the blues Polyommatinae, Marsh Fritillary *Euphydryas aurinia*, and Marbled White *Melanargia galathea*.

Whether it is accurate to describe such Lepidoptera as colonial depends on how strictly the concept is defined. Caterpillars from the same batch of eggs sometimes stay together (Plate 45), but a brood is hardly a colony. Many birds and some animals are colonial, but here there is always some form of social co-operation and interaction between members, involving at least mutual warning and defence of the colony against predators. Among the insects, the termites, ants, and some bees and wasps actively co-operate in this way, and individuals often forage not for themselves but to provide for the colony as a whole. They too must qualify as colonial.

Such mutual interdependence seems absent in the Lepidoptera. Perhaps apparently colonial species are ones that need to maintain a relatively high density of individuals for successful breeding, and can only achieve this in the most favourable spots. However, they might deserve to be called colonial if there is some mechanism for preventing individuals from straying. This is a possibility. I have watched male Small Blue *Cupido minimus* butterflies wander several metres away from their colony but invariably turn back and rejoin it. The New Forest Burnet *Zygaena viciae* is quite mobile (given suitable weather) within its small Argyll site, crossing obvious breaks in habitat, yet never seems to stray beyond invisible boundaries into apparently suitable neighbouring areas. It is feasible that pheromones given off by the other members of a colony serve to hold it together.

Warningly coloured and distasteful (aposematic) species like the burnets and Scarlet Tiger might even qualify as genuinely colonial on the grounds of communal defence. If individuals were scattered at a lower density, this would take them into the territories of many more predators such as birds, each of which might need to learn by trial and error that the insect was distasteful. In a colony, however, not only is the warning message reinforced by numbers, but there are fewer nearby predators to be educated.

What is a 'successful' species?

Moths show every permutation between abundant and ubiquitous species to scarce and very local ones. They range from generalists able to eat a wide range of food plants and thrive almost anywhere, to specialists that are restricted to a single food plant, a certain climate, or a precise habitat. Can we say that some species are more successful than others? First there is the question of defining what is meant by successful. Should we base the qualification on an extensive geographical distribution, the ability to use a wide range of habitats, or on size of population? By all these counts, humans would qualify as successful, but any specialised, localised, low-density or niche species automatically would not. This seems too restrictive. From the point of view of an individual moth attempting to pass on its genes, it does not matter how abundant or widespread its species is. Granted, the population must be large enough for it to find a mate, to maintain a varied gene pool, and to avoid extinction in a bad year. Apart from that, there might even be some advantages in not being numerous enough to attract the attention of predators. There are certainly many moths that are neither common nor scarce, are not subject to any great fluctuations, but just carry on unobtrusively year after year. Surely we should regard them as successful species in their own way? There are no such problems, however, in defining unsuccessful ones – they become extinct!

Losses to the British list

About 65 moths (macros and micros) are thought to have become extinct in Britain and Ireland since reliable records began around 150 years ago. Unfortunately, the picture is not always clear. The older authors did not realise how many species are only migrants to Britain, believing them to be indigenous but scarce. As well as honest confusion, there was subterfuge. In the nineteenth century, the buying and selling of moths and butterflies was a considerable business, and it was in the financial interests of those involved to claim a British origin for the specimens or livestock they were offering for sale. Certain dealers, such as Plastead (Allan, 1943), were notorious for importing desirable species from the Continent and selling them to unwary collectors with false data and fraudulent labelling. It was the lepidopteran equivalent of ornithology's Hastings Rarities. As genuine new species were continually being added to the British list at this time, even the experts were sometimes fooled. Although most of these dubious records have now been discounted, there remains the tantalising possibility that a few were valid.

The Pease Blossom *Periphanes delphinii* is a beautiful moth that once may have been resident in southern English gardens, feeding like the Golden Plusia *Polychrysia moneta* on cultivated Monk's-hood and Larkspur – or perhaps that is wishful thinking.

The second difficulty is that of proving a negative: how much evidence is required before we can say whether a species is extinct? Even if we are sure that it no longer occurs at any of its former sites, perhaps it still survives at others yet unknown. There are many instances of the rediscovery of moths long assumed to have been lost. The Rosy Marsh Moth *Coenophila subrosea* was believed extinct in Britain after the draining of the Fens was completed around 1850, only to turn up in a Welsh bog in 1965. It is just possible that this was a re-colonisation following migration. However, such an explanation would not fit the sedentary New Forest Burnet, which died out in Hampshire in 1927 but was rediscovered in Argyll in 1963, making a mockery of its English name. The microlepidoptera are even easier to overlook. *Ethmia pyrausta* owed its place on the British list to a single example captured in Sutherland in 1853, but all subsequent searches in that area were unsuccessful. Then, in 1996, two were found at over 1,000 metres above sea level in the Cairngorms (Young and Smith, 1997). Ironically, they were not caught by a lepidopterist, but fell into pitfall traps set for crane flies.

It is instructive to look at the regional pattern of losses within Britain. Figure 2 confines itself to the macro-moths, as there is less uncertainty about past and present status in this group. The map shows that of 25 former residents presumed gone from the British Isles, 18 had their strongholds in eastern or south-eastern England. Otherwise, remarkably few moths have died out, considering the scale of the changes that have taken place in our towns and countryside during the past two centuries.

Habitat destruction is certainly the chief factor in most of the losses from eastern England. Drainage of the Fens for agriculture was catastrophic for the specialist moths that lived there. The Many-lined *Costaconvexa polygrammata* disappeared in about 1850 when Burwell and Wicken fens were drained, and the Rosy Marsh Moth, Gypsy Moth and Reed Tussock *Laelia coenosa* likewise succumbed when their haunts were destroyed. The lowering of the water table may have caused subtle changes to the climate of the whole region. Species associated more with the margins of Fens, like the Marsh Dagger *Acronicta strigosa* and the Orache Moth *Trachea atriplicis*, lingered on into the twentieth century before dying out.

East Anglia's other unique habitat, the dry and sandy Breck, also supports a characteristic fauna of rare moths. Typically, their food plants are weeds of waste places and the edges of arable fields, themselves very vulnerable to modern agriculture's efficiency. Afforestation with conifers has caused further loss of habitat. Consequently, two of Breckland's most charismatic

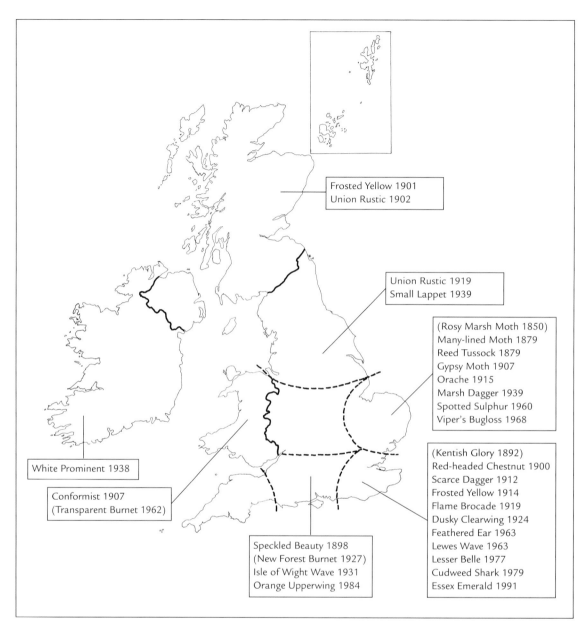

Frosted Yellow 1901
Union Rustic 1902

Union Rustic 1919
Small Lappet 1939

(Rosy Marsh Moth 1850)
Many-lined Moth 1879
Reed Tussock 1879
Gypsy Moth 1907
Orache 1915
Marsh Dagger 1939
Spotted Sulphur 1960
Viper's Bugloss 1968

(Kentish Glory 1892)
Red-headed Chestnut 1900
Scarce Dagger 1912
Frosted Yellow 1914
Flame Brocade 1919
Dusky Clearwing 1924
Feathered Ear 1963
Lewes Wave 1963
Lesser Belle 1977
Cudweed Shark 1979
Essex Emerald 1991

White Prominent 1938

Conformist 1907
(Transparent Burnet 1962)

Speckled Beauty 1898
(New Forest Burnet 1927)
Isle of Wight Wave 1931
Orange Upperwing 1984

Figure 2: The regional pattern of major losses among macro-moths in the British Isles since *c.* 1850.
Species in brackets represent regional losses only.

species, the Spotted Sulphur and the Viper's Bugloss *Hadena irregularis* have slipped away in recent years (Haggett, 1998), as unremarked at the time as the last autumn Swallow.

In south-east England, habitat deterioration has been more gradual and piecemeal, but still there are convincing instances where it has caused the loss of a species. The Lewes Wave began to decline at its Sussex location in the 1940s, when wartime cultivation started to affect the heathland habitat. Then most of the site was planted with conifers, and Bracken invaded the rest. The moth was last seen in 1963. A disastrous summer fire in 1967 put paid to any hope that it could have survived. Pratt (1981) documents this sorry story. He considers the moth's demise could have been avoided, given better co-operation between foresters and conservationists.

However, valiant efforts failed to save the Essex Emerald *Antonechloris smaragdaria* from extinction. Always restricted to salt marshes in Essex and Kent where the caterpillar fed on Sea Wormwood, it probably suffered from fragmentation of its habitat, isolating colonies which then dwindled until the last died out in 1991. A captive population had been established in 1987, but all attempts at reintroduction were unsuccessful. Inbreeding affected the vitality of the captive stock (this might also have been a factor in the wild), and it too succumbed in 1996 (Waring, 1990 and 1996a).

Other instances are less clear cut, and moths have been lost without any obvious destruction of their habitat. Many parts of its former range in Sussex still appear suitable for the Flame Brocade *Trigonophora flammea*. The Isle of Wight Wave *Idaea humiliata* has not been seen on the Freshwater undercliffs since 1931. We might blame the supposed general, insidious degradation of the countryside, or pollution, or collectors, but that would be lazy. It would also be symptomatic of modern society's urge to allocate culpability whenever something unfortunate happens. Farmers, industrialists and governments make easy targets.

I believe it is arrogant as well as wrong of humans to assume that they themselves are responsible for each and every loss of a species – or could somehow have prevented this happening. The climate of eastern and south-eastern England is closest to that of the Continent: dry, with sunny summers and relatively cold winters. Many moths more at home in mainland Europe have been able to gain a precarious foothold here, often in very restricted areas that offer a particularly suitable habitat. For instance, several moths have been able to establish a breeding population only on the shingle at Dungeness, like the Kentish Plover once did. Sooner or later, it is inevitable that even minor adverse changes such as a run of cool summers, a build-up of parasitoids, or vegetation succession, will tip the balance, causing them to die out. Probably we should regard most as temporary residents anyway: even a hundred years is short on the biological timescale.

For several of these species, there is already a clear pattern of temporary residence, with extinction followed by re-establishment, presumably by migrants. The Rest Harrow *Aplasta ononaria*, Sub-angled Wave *Scopula nigropunctata* and the Scarce Chocolate-tip *Clostera anachoreta* seem to have come and gone several times in the past 150 years. It is no coincidence that many of our scarce, very local, threatened or extinct species belong to the subfamily Sterrhinae, the waves. This is predominantly a southern group. Of the 188 species known to occur in Europe, 139 (74 per cent) are found in Spain, but only 36 (19 per cent) in Britain and 27 (14 per cent) in Norway, according to Karsholt and Razowski (1996). Just seven of the British species are graded as common by Waring (1993), the others being accorded local, notable, or Red Data Book status, if they are not already extinct.

There have not been many national losses among the macro-moths confined to central, western and northern Britain. Partly, this is because there are fewer species in that category. For example, Warwickshire has no Red Data Book species in its list of 588 macro-moths (Brown, 1997). All south-west England's special rarities of coastal cliffs are surviving to date, but in south Wales The Conformist *Lithophane furcifera* has almost certainly been lost. The Small Lappet *Phyllodesma ilicifolia* is the chief casualty in northern England. It was always rare, and hard to obtain, so there are hopes it might still be there. Most were found as caterpillars by the gatherers of wild bilberries who supplied the confectionery trade. This practice continued at least into the 1960s in northern England, where bilberry tarts were a local delicacy in my boyhood, but caterpillars had long since ceased to be reported.

Plate 46: Scarce Silver Y *Syngrapha interrogationis*. Fortunately this moth is unquestionably commoner than its English name suggests: like most northern upland species, it seems to be doing well. (Banffshire).

Perhaps dealers were no longer offering financial incentives for any found.

Scotland does have a characteristic fauna of northern species, some of which are local. About two dozen macro-moths are either confined to that country (as regards Britain) or have their stronghold there. None has yet been lost, become noticeably scarcer, or suffered a serious reduction in range. Nor do any species except the New Forest Burnet appear to be endangered at present: like Scotland's butterflies, most seem to be thriving. The two most significant losses involve moths that also disappeared from England in roughly the same period. One was the Union Rustic *Eremobina pabulatricula*, which was last recorded from central Scotland in 1902, and from its best-known sites in Yorkshire and Lincolnshire in 1919. The reasons for its decline are obscure, but it appears to be local and rare elsewhere in Europe. The other moth is the Frosted Yellow, a species with a rather restricted European distribution. The mystery is what it was doing in northern Scotland at all, so far beyond the rest of its range. Ford (1955) illustrates a specimen from inland Ross-shire, and there are also records from Perthshire, but the last Scottish sightings were in 1901.

The caterpillar feeds on Broom, so does not lack for food plant, and there is just a chance the moth will be rediscovered. Like *Ethmia pyrausta*, it is day-flying, and therefore not susceptible to light traps, the predominant method of collecting today. However, there is little room for doubt that the Frosted Yellow really has gone from its English sites in Kent, Essex and East Anglia, where it was last seen in 1914.

The Notodontidae, by contrast, are strongly attracted to ultraviolet light. Consequently, had the White Prominent *Leucodonta bicoloria* still occurred in the unspoilt birch woods of Killarney or elsewhere in southern Ireland, it would surely have been found by those who have gone to look for it. Discovered in 1859, it was last seen in 1938. This striking moth is widespread and common on the Continent, and there is no obvious reason why it is no longer found here. In England, just six examples were taken (it is said) in Burnt Wood, Staffordshire, between 1861 and 1865. However, Tutt (1902) himself had seen over 20 different specimens bearing that labelling subsequently sold at auction! This illustrates the scale of fraud prevalent at that time.

Firm evidence that the moth was correctly identified in the first place, or that it was truly resident, is even harder to obtain for the microlepidoptera. Appendix 2 lists 30 species that, on the balance of probabilities, were former residents that have since been lost. There is difficulty in evaluating whether a lack of recent sightings means that the moth has died out, or has simply been overlooked: the tiny *Leucoptera sinuella* may well survive although it has not been reported from its Speyside haunts since 1950. However, the pattern is similar to that of macro-moths: most of the lost species were confined to southern or eastern England.

To summarise, the evidence suggests that habitat destruction has had its greatest impact on Lepidoptera in East Anglia, where the draining of the Fens and fourfold reduction in area of the Breck has wiped out some of the long-established, specialised moths that lived there. Except for cases like that of the Lewes Wave, loss or degradation of habitat is not so obvious a cause of extinction in south-east and southern England. Most of the moths involved were species at the edge of their European range. Vulnerable to the slightest adverse factors, especially climatic ones, they come and go. In contrast, very few of the moths confined to northern England, Scotland or Ireland have become extinct, and for those that have, destruction of habitat is not the likely cause.

Gains to the British list

If the above analysis is correct, we should expect southern England in particular to gain species as well as lose them. Indeed, gains and losses should roughly cancel each other out. This does seem to be the case. Figure 3 maps the macro-moths gained by colonisation since 1870; before then, the

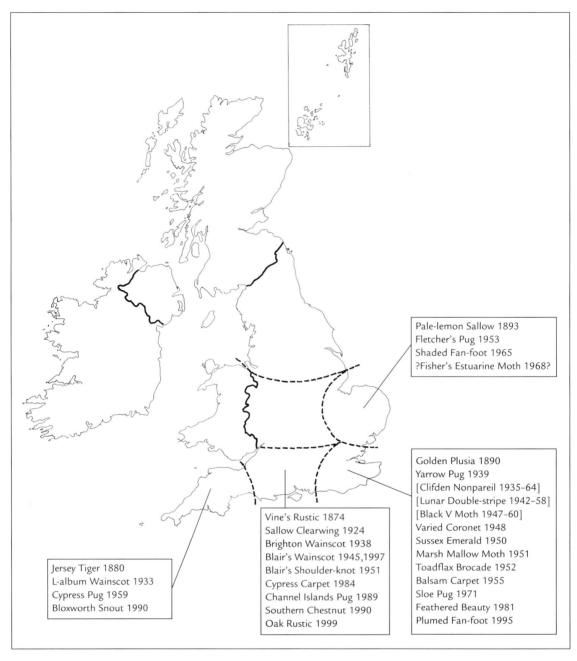

Pale-lemon Sallow 1893
Fletcher's Pug 1953
Shaded Fan-foot 1965
?Fisher's Estuarine Moth 1968?

Golden Plusia 1890
Yarrow Pug 1939
[Clifden Nonpareil 1935–64]
[Lunar Double-stripe 1942–58]
[Black V Moth 1947–60]
Varied Coronet 1948
Sussex Emerald 1950
Marsh Mallow Moth 1951
Toadflax Brocade 1952
Balsam Carpet 1955
Sloe Pug 1971
Feathered Beauty 1981
Plumed Fan-foot 1995

Vine's Rustic 1874
Sallow Clearwing 1924
Brighton Wainscot 1938
Blair's Wainscot 1945,1997
Blair's Shoulder-knot 1951
Cypress Carpet 1984
Channel Islands Pug 1989
Southern Chestnut 1990
Oak Rustic 1999

Jersey Tiger 1880
L-album Wainscot 1933
Cypress Pug 1959
Bloxworth Snout 1990

Figure 3: The regional pattern of major gains of macro-moths in the British Isles since *c.* 1870. Species in brackets were temporary colonists only.

country's moth fauna was not well enough known to recognise new arrivals. As always, there may be some dispute about which species qualify. On the balance of probabilities, I have included several that are not generally recognised as newcomers. The Pale-lemon Sallow *Xanthia ocellaris* is one, as South's (1907–09) account strongly suggests colonisation. The first were taken in the London area in 1893, yet by 1907 there were reports from Kent, Suffolk, Norfolk and Cambridgeshire, roughly the area it occupies today. It was unlikely to have been overlooked until then, as a similar species, the

Dusky-lemon Sallow *X. gilvago*, had several times been mistaken for it. Ford (1955) considered the Brighton Wainscot *Oria musculosa* to be native, after EA Cockayne and B Kettlewell found it abundantly in 1938 by following the harvester in wheat fields near Salisbury. Previously, the moth had been known only as an extremely rare immigrant. Would such a distinctive species, if resident, have gone unnoticed in the age when grain was reaped and bound by hand? Nor do I consider that the Sloe Pug *Chloroclystis chloerata*, first recorded in 1971, had long been present. By 1997, there were records from almost two hundred 10 km squares (Waring, 1997), as far north as Westmorland. It is hard to believe that earlier workers would miss so widespread a moth, or that no wrongly identified specimens would be found in old collections. Likewise, if the Sallow Clearwing *Synanthedon flaviventris* had been here long before 1924, it would probably have been found by those working sallows for the related Red-tipped Clearwing *S. formicaeformis*.

The expertise and diligence of Victorian lepidopterists should never be underestimated. They lacked the powerful mercury vapour traps generally used today. In a way, this was an advantage: it meant that they were forced to employ a much wider range of methods to find moths. Far more effort went into looking for the early stages. Adult moths were found by daytime searching, netted at dusk, lured to sugar, or taken at natural attractions such as sallow blossom and ivy bloom. Because scarce species were saleable, the profit motive was combined with scientific curiosity, a potent brew.

As a result, Britain's moth fauna was very well known by the end of the nineteenth century, even in the more remote areas. Remarkably, only one resident macro-moth new to Britain has been discovered in Scotland during the twentieth century: the Slender Scotch Burnet *Zygaena loti* in 1907. Wales has provided The Silurian *Eriopygodes imbecilla* (1972), and western Ireland two species, the Burren Green *Calamia tridens* (1949) and the Irish Annulet *Odontognophos dumetata* (1991), both of which differ enough from the European type to be accorded subspecies status, proof that

Plate 47: Small Dotted Footman *Pelosia obtusa*. Although only added to the British list in 1961, this very local fenland species is probably one of our few resident macros that the Victorian collectors overlooked. (Norfolk).

they are long established. In England, it is quite plausible that even the Victorians could have missed Fletcher's Pug *Eupithecia egenaria* (Plate 57) and the Small Dotted Footman *Pelosia obtusa* (Plate 47). Both are small, drab and very local species, but Haggett (personal communication) still considers that the former is probably a newcomer.

Among the microlepidoptera, nearly 60 species have strong claims to be regarded as recent colonists, outnumbering extinctions by 2:1. They are listed in Appendix 2. Together with the macro-moths, they tend to fall into two groups. First, there are those that have never lacked for food plants here, but apparently find the climate unsuitable for long-term residency. Generally, they have not spread very far after arrival. Secondly, there are species previously absent because their food plant does not occur here naturally. They feed on exotic trees (especially conifers), shrubs and herbaceous garden plants, and in some cases may have travelled with them as eggs or caterpillars. Often, they have spread rapidly once they have arrived, some even reaching Scotland. Agassiz (1996) classes these as invading species, and maps the dramatic rate of spread of moths like the Golden Plusia (Delphinium and Larkspur), the Varied Coronet *Hadena compta* (Sweet William) and the leaf-miner *Phyllonorycter leucographella* (Firethorn). He separates these successful invaders from 'failed establishments' that have only been able to gain a small foothold in Britain, but this distinction is very arbitrary. For instance, he classes the L-album Wainscot *Mythimna l-album* as a failure despite its spread along the south coast as far as Kent since its arrival in Cornwall in the 1930s.

Three of the newcomers are associated with the Leyland Cypress, that vigorous but sometimes hated hedging tree of suburban gardens. Arising from hybrid stock in 1888, there are now estimated to be about 55 million of these plants in Britain. This is a considerable food resource, but Freyer's Pug *Eupithecia intricata arceuthata* seems to be the only native moth able to use it, switching from its normal food plant, Juniper. As a result, this moth has become much commoner and more widespread. Remarkably, three moths with a largely Mediterranean distribution have colonised southern England, feeding on this and other introduced cypresses. First, in 1951, was Blair's Shoulder-knot *Lithophane leautieri*, which has now spread into Wales and northern England as far as Westmorland and may soon reach Scotland. Next came the Cypress Pug *Eupithecia phoeniceata* (Plate 48) in 1959, our only pug that overwinters as a caterpillar. The Cypress Carpet *Thera cupressata*, first recorded in 1984, is the latest to try and establish itself. Gardens are sheltered, and have a drier and warmer microclimate than the open countryside, which may have helped these and other more southern species to survive here.

In almost every case, the new coloniser had never previously been recorded in Britain as a migrant, or if so, only very rarely. This is often used as an argument that the species might have been resident here all along,

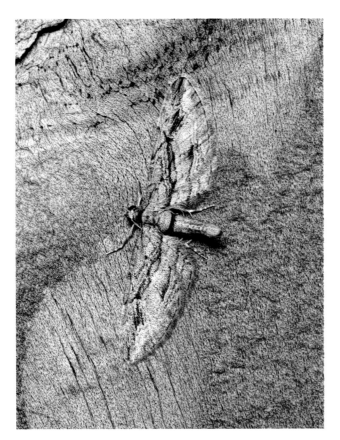

Plate 48: The Cypress Pug *Eupithecia phoeniceata* is one of several Mediterranean moths to have become established on exotic garden cypresses. It is slowly spreading in southern England. (Sussex).

unnoticed. However, it is quite understandable that most newcomers fall into this category. The commoner and more regular migrants have had numerous chances to settle, yet have failed to do so. If they could colonise Britain, they would have done so long ago. It is precisely those species that hardly ever get the opportunity to colonise that are most likely to take advantage of it if they do. So it should not have been a surprise when the L-album Wainscot became established on the south coast, rather than any of the much less scarce migrants in the same genus. The American Wainscot *Mythimna unipuncta*, White-point *M. albipuncta* and the Delicate *M. vitellina* have dozens if not hundreds of chances to colonise in every normal year. Clearly, the climate does not suit them. Similarly, it is a forlorn hope that the Camberwell Beauty *Nymphalis antiopa* butterfly will ever become resident in Britain unless the climate changes or the butterfly itself becomes more tolerant.

One of the most recent colonists is the Southern Chestnut *Agrochola haematidea*, discovered on a Sussex heath in 1990 and found in the New Forest in 1996. This is another western Mediterranean species, but the caterpillar's food plant is Bell Heather (Haggett and Smith, 1993). Again, the moth is not a known migrant, and there were suggestions that it had previously been overlooked even in such a well-worked part of the country. However, it may be significant that three other members of its genus, also not regarded as migratory, have been recorded from oil rigs in the North Sea. They are the Brick *A. circellaris*, Yellow-line Quaker *A. macilenta*, and Brown-spot Pinion *A. litura* (North Sea Bird Club, unpublished files). Incidentally, this group is another example where the scientific names show kinship, but the common names do not.

Often, it is those species with no realistic chance of reaching a country by their own efforts that prove to be the most invasive colonisers when artificially transported there, either accidentally or deliberately. Nostalgic introductions of common British songbirds to the Antipodes and North America have often been all too successful, to the detriment of native species. Even small islands far off the coast of New Zealand now have breeding Dunnocks,

Blackbirds, Song Thrushes, Starlings, Redpolls and Chaffinches (Campbell and Lack, 1985). The traffic has been mainly one way, but the Australian tortricid moth *Epiphyas postvittana* has become well established in the southern part of Britain since the first breeding population was found in Cornwall in 1936. Probably it was introduced with imported apples, being an orchard pest in its original home. Fortunately, here it prefers the garden hedging shrub *Euonymus japonicus*.

As far as we can tell, none of Britain's recent colonists has had adverse effects on the native species. There is no moth equivalent of the Grey Squirrel or Signal Crayfish. Many of the most successful invaders feed upon introduced plants, so do not compete directly with indigenous moths. Of course, there may be subtler effects of which we are not aware: the newcomers might act as hosts for shared parasitoids, disrupting an existing balance, or provide a reservoir for diseases. There is no evidence of this yet. Even the introduction of almost cosmopolitan pests of cultivated plants or stored produce have proved little more than minor nuisances in modern Britain. Thus virtually all our gains can (so far) be regarded unequivocally as welcome additions to Britain's moth fauna, compensating us to a great extent for the species we have lost.

In purely numerical terms, there has been a net gain of moths over the past 150 years. This is surprising, given the widely held belief that the countryside is becoming ever more despoiled and its wildlife endangered. Yet the argument that Britain's moth fauna has been impoverished can only be sustained by making highly subjective value judgements about the relative worth of the species gained and lost. Does the arrival of the Sussex Emerald fully compensate for the loss of the Essex Emerald? (Of course it would be nice if we still had both!) Is it possible to say whether the Shaded Fan-foot *Herminia tarsicrinalis* is a fair swap for the Lesser Belle *Colobochyla salicalis*, the Varied Coronet for the Viper's Bugloss? In the case of the last pair, the exchange can be considered a poor deal, because the lost species was much more local in Europe than the one gained. However, the three cypress feeders mentioned earlier also have rather restricted European distributions, so in conservation terms they might be seen as valuable assets.

Changes in range and abundance

It is too simplistic to look only at gains and losses. Even if Britain still has about the same number of moth species, ranges might have contracted, and numbers of individuals decreased. Again, the evidence is far from conclusive. It is easy to find examples where this is so, but equally easy to point to moths that have become much commoner and more widespread. At the beginning of the twentieth century, Tutt (1902) considered the Brown-tail practically extinct. In recent years, local authorities on the south coast have been forced

Plate 49: Dusky Sallow *Eremobia ochroleuca*. Not all moths have declined: relatively scarce a century ago, this dry grassland species is now more widespread, and sometimes abundant on chalk downland. (Sussex).

to destroy larval webs as a public health hazard because of the caterpillars' irritant hairs. He described the Chocolate-tip *Clostera curtula* as much scarcer than its relative, the Small Chocolate-tip *C. pigra*, with stock even being imported from abroad to supply collectors. It is by far the commoner species today. He regarded the Barred Sallow *Xanthia aurago* as very local, now, it occurs over most of England and Wales. Kent is the only county Tutt mentions for Dusky Sallow *Eremobia ochroleuca* (Plate 49); it has since spread as far north as Yorkshire.

It is often argued that the availability of mercury vapour light traps has shown some species to be much commoner than once thought, and there is some truth in this. The treetop-dwelling Alder Moth is often mentioned in this context. However, it can hardly be the explanation when species of similar habitat and behaviour are involved. Tutt (1902) considered the Brown-veined Wainscot *Archanara dissoluta* to be one of our rarest moths, found only in Essex and Cambridgeshire, yet stated that its congener the Twin-spotted Wainscot *A. geminipuncta* abounded in every southern reed bed. Webb's Wainscot *A. sparganii* was also a rare species then, recorded only from marshes between Hythe and Deal, as South (1907–09) agreed. Today, all three species have a very similar distribution and status, mainly coastal and extending at least as far as south Wales and Lincolnshire.

The extent of the subsequent spread of these two wainscots strongly suggests that they might have been recent colonists when first reported in Britain. The same might be said of the Kent Black Arches *Meganola albula* (Plate 128), once very local but now found over much of southern England. The Pine Hawk-moth (Plate 50) was so rare in South's time that he lists most sightings individually. Nearly all were from Suffolk. It has since colonised so much of southern and eastern England that it no longer qualifies as a nationally notable species, and in some places is the commonest hawk-moth (Webb, 1984). Like other moths associated with conifers, and the Siskin, it has benefited from the very forestry plantations that have destroyed the habitat of other scarce species. The Dwarf Pug *Eupithecia tantillaria* is yet another probable colonist, as it could not occur here until its food plants, spruce and

Plate 50: Pine Hawk-moth *Hyloicus pinastri*. Like other species associated with conifers, this fine moth has prospered in southern Britain as forestry plantations have matured, and is still extending its range. (Norfolk).

fir, were widely planted. In South's time it was local, and only in southern England, but is now found throughout Britain. Similarly, the Larch Pug *E. lariciata* was unknown here until 1862, but now is equally widespread.

The best example of a resident moth, not a recent arrival, which has undergone a dramatic expansion in range and numbers is the Waved Black. Unusually for a macro-moth, its caterpillar feeds on bracket fungi growing on dead wood. Once a great rarity, found only in the London area, it has now been recorded from over a hundred 10 km squares and may still be spreading northwards. Changes in woodland management are thought to be behind the increase. In the days before central heating, fallen timber was a valuable resource. Yonder peasant would have gathered the branches for his winter fuel long before any *Coriolus versicolor* began to grow on them! Today, such branches are left to rot.

Inevitably, there are also species that have declined. The Narrow-bordered Bee Hawk-moth (Plate 16), once found locally throughout Britain and Ireland, is now very sparsely distributed. Like other diurnal moths, it is probably under-recorded because of the modern over-reliance on mercury vapour light traps, but the decline is genuine. Drainage of wetlands, where its food plant Devil's-bit Scabious grows, is often blamed, but it is hard to see how this could apply in Scotland as ample apparently suitable habitat remains. However, there are encouraging signs of a recovery taking place in the Scottish Highlands at least, with many recent sightings in Moray and Inverness-shire where the moth had not been seen for years, and pupae even found in gardens (S Moran, personal communication). It has also been rediscovered in East Anglia (Musgrove and Armitage, 2000).

The Sword-grass (Plate 101) has been retracting its range northwards in Britain for at least the last hundred years, for reasons quite unknown, disappearing from south-east England first. Now, it is thought to be resident only in Scotland. The Pale Shining Brown *Polia bombycina* has gone the other way: once found in Scotland, it is one of very few moths that country has lost, having retracted its range towards south-east England. Caution is needed when assessing the old records. The moth itself is distinctive, but a common northern species, the Glaucous Shears *Papestra biren*, was also wrongly known as *P. bombycina*. Reports of '*P. bombycina*' from Scottish localities are causing confusion to this day! However, actual specimens of the Pale Shining Brown taken in Perthshire in the nineteenth century are housed in the Scottish national collection at Edinburgh (K Bland, personal communication). Such firm evidence is not available for the Large Nutmeg *Apamea anceps*, given by the old authors as extending through eastern Scotland as far as Moray. There are no modern records, and no voucher specimens exist, so it may have been confused with similar species in the same genus.

It is less easy to determine whether a moth has maintained all its previous range, but declined in abundance. Earlier writers make subjective assessments of status, but rarely give actual numbers. When they do, it is often uncertain whether they are reporting unusual extremes, or the normal population level. However, it seems likely that four moths long associated with garden currants and gooseberries are much less common than they used to be. Perhaps fewer fruit bushes are grown, or more effective insecticides are used. I suspect, too, that modern gardeners are more willing to share their crop with the birds they actively encourage into their gardens. Fruit bushes are less regularly caged or netted, thus exposing the early moth stages to

predation. That should not have affected the highly distasteful Magpie Moth,
but tits peck at the pupae of the Currant
Clearwing *Synanthedon tipuliformis*, still
listed as a pest in horticultural hand-
books but now a nationally notable
species. Both the Spinach *Eulithis mellinata* (named after the wrong plant) and
the V-moth *Macaria wauaria* (Plate 51)
are hard to find these days; the latter was
described as an injurious species by Ford
(1955), and was a common boyhood
moth for me, though we had but a single Gooseberry bush. I have seen one in
the past 20 years.

One of the most dramatic changes in
status of a moth that has not yet become
extinct in Britain is shown by the Four-
spotted *Tyta luctuosa*. It is an attractive
day-flying species, formerly found over
large parts of England south and east of a line from the Humber to the
Severn, especially on the chalk. Tutt (1902) confidently states 'most coun-
try-lovers must have seen it'. Today, few lepidopterists have. Its food plant,
Field Bindweed, remains common enough.

No mystery surrounds the massive declines suffered by some elm specialists in the aftermath of Dutch Elm Disease. The White-letter Hairstreak
Satyrium w-album butterfly was badly affected, while among the moths,
White-spotted Pinion *Cosmia diffinis* is the best-documented case. Waring
(1996b) maps all records before and after 1980. Much less attention was
paid to the Clouded Magpie because it was not considered a nationally
notable species, but I suspect it too has been severely hit, according to reports
from recorders in counties where once it was common. This shows the danger of relying on a single food plant.

However, the overall pattern remains a balanced one of gains and losses. For every moth that has become scarcer and more localised, there is
another that has extended its range and increased in numbers. Granted, certain communities of moths have done badly – Breckland and chalk downland species for instance – following obvious habitat reduction. But, thanks
to forestry and garden centres, moths associated with conifers have prospered. Not all changes have been bad. Nevertheless, few would contest the
accepted dogma that the countryside, and farmland in particular, has become
poorer for wildlife during the twentieth century. Yet moths have managed
to cope surprisingly well – especially when compared with their close rela-

Plate 51: Like several moths associated with garden currants and gooseberries, the V-moth *Macaria wauaria* seems less frequent today, and is certainly not a pest species as described in the past. (Norfolk).

tives. Britain has lost five resident species of butterfly in recorded history and gained none, unless the Geranium Bronze *Cacyreus marshalli*, a recent adventive from South Africa, proves able to survive our winters. Otherwise, the Black-veined White *Aporia crataegi*, Large Copper *Lycaena dispar*, Mazarine Blue *Cyaniris semiargus*, Large Blue *Maculinea arion* and Large Tortoiseshell *Aglais polychloros* are uncompensated losses (though two have been reintroduced), as is the Chequered Skipper *Carterocephalus palaemon* from England. Although none of the fritillary butterflies has yet become extinct here, all have undergone serious declines.

In the Netherlands, the situation is even worse. Of 70 species of previously resident butterflies, 17 (nearly a quarter) are now extinct, and a further 28 are graded from vulnerable to critically endangered. Only 23 species are considered safe (Maes and van Sway, 1997). Yet the Netherlands shows a net gain of moths, at least among the microlepidoptera (Kuchlein and Ellis, 1997).

Why are moths doing better than butterflies?

Moths seem more able to cope with modern changes to their environment. This raises the old question: what is the difference between a butterfly and a moth? My own hypothesis is that moths can survive at a lower density and in more fragmented habitats because of the way they find their mate. Moths use pheromones, which are effective over long distances downwind (probably at least several hundred metres). Butterflies find their mates mainly by sight (Ford, 1945). For this to work, the minimum density of individuals in a population must be relatively high. Many butterflies will not cross unsuitable ground, whereas a male moth will follow a pheromone trail whatever the terrain. A virgin female Emperor Moth put out in a city backyard has a good chance of attracting a male (Plate 52), but a female Meadow Brown butterfly is very unlikely to do so, even if that species is plentiful in the surrounding countryside.

To locate the food plant for egg-laying, female butterflies are thought to use sight followed by touch or 'taste', involving special sense organs on the tarsi of the last pair of legs (Ford, 1945). It is unlikely that night-flying moths use sight, so perhaps they employ scent for this purpose too. If so, it would again give them an advantage over butterflies where the appropriate habitat was discontinuous and the food plant widely scattered. Spence (1991) relates how a Foxglove appeared at Spurn in 1988, the first for the peninsula since records began, and a Foxglove Pug was caught the same summer. This was too soon for the moth to have been raised on that plant, as the caterpillar eats only the flowers, then overwinters as a pupa. Coincidence, perhaps, but the first Foxglove Pug for my own garden in Banffshire also turned up the same summer that my introduced Foxgloves flowered. No others grew in the area.

Size might be a further factor. No butterflies are as small as the vast majority of micro-moths. Perhaps the latter can thrive in very small patches of suitable habitat – even a single plant – whereas butterflies need larger, continuous areas that are now in such short supply.

Hitching a lift

There is one way in which human intervention has benefited certain moths immensely, enabling them to extend their range and increase in numbers far beyond what would otherwise have been possible. The Atlantic Ocean is a formidable barrier. Given the right weather systems, it can be crossed from west to east by non-marine species, as shown by the small American passerines that turn up in the Scillies every autumn, the Monarch *Danaus plexippus* butterfly, and the Green Darner *Anax junius* dragonfly (Davey, 1999). The reverse crossing, against the prevailing westerlies, can safely be described as impossible for an insect under its own power.

However, a surprising number of European moths have become established in North America, almost certainly as introductions through human agency. The information is piecemeal, and I was unable to trace, either in Britain or America, any collated list of the species involved. Accordingly, my own compilation is given in Appendix 3. It is certainly very incomplete, and one or two species might be wrongly included. It should only be regarded as a framework for further work. Even so, it includes 76 moths and

Plate 52: One of everybody's favourite moths, the male Emperor *Saturnia pavonia* is renowned for its ability to locate females by scent over long distances, using its complex antennae. The purpose of the eye markings is less clear: warningly coloured species normally do not show such subtle patterns and tones. (Sussex).

Plate 53: Cinnabar *Tyria jacobaeae* caterpillars feed openly, relying on their striking livery to proclaim their toxicity due to poisonous alkaloids sequestered from their food plant, ragwort. Various attempts have been made to use them for control of this noxious weed. (Kincardineshire).

2 butterflies on the British list that have apparently been introduced to North America.

Most of the moths involved belong to the microlepidoptera. As a rule, it is only those that have drawn attention to themselves by becoming pests in their new home that are documented in the literature. Some moths (or rather their caterpillars) have been given common names, such as Apple Leaf Skeletonizer *Chloreutis pariana* and Omnivorous Leaf-tier *Cnephasia longana*, reflecting an American tendency to see moths as enemies rather than as wildlife. After it was introduced, our Rosy Rustic became their Potato Stem Borer! When I excused myself one evening to go and put out the moth trap, my visiting father-in-law from New York State exclaimed 'Gee! I never realised moths were such a Big Problem here in Britain.' At his Long Island home, specialist firms routinely come out to spray whole gardens, including ornamental trees 40 or 50 feet high, against caterpillars.

Some of the introductions are indeed major pests. One Leopold Trouvelot deliberately took the Gypsy Moth to Massachusetts in 1868, hoping to use it for commercial silk production (Covell, 1984). Unfortunately, it escaped the next year. Nearly every North American tree, deciduous or coniferous, is attacked, and the caterpillars can do enough damage to kill the trees. The commercial cost has been incalculable, in spite of control attempts involving synthesised pheromone lures or bacteria (Young, 1997), while the range of the moth is still increasing. Another of the tussock moths, the Brown-tail, became a major pest on the East Coast after its introduction in 1897 (South, 1907–09), but is now confined to seashore habitats on islands.

Other moths have been purposely introduced to North America to attempt biological control of weeds. The Cinnabar *Tyria jacobaeae* (Plate 53) is one, intended to control ragwort (Heath and Emmet, 1979). Many British farmers must have tried this tactic, after seeing the effect on that plant when caterpillars are present in large numbers, as happens at some coastal sites. A Sussex farmer of my acquaintance brought home dozens, and released them on his land with great hopes, not knowing that the moth was already present there at a low density. More success attended the release of the Spurge Hawk in western Canada to control *Euphorbia* weeds. At least, the moth is now established there (Watson and Whalley, 1975), but of course no species can eliminate its food plant. Otherwise, most of the introductions are clearly accidental. The colonisation of North America by Lepidoptera from Europe mirrors the pattern of human immigration. For economic and sentimental reasons, these settlers took with them familiar plants from their native land, both food and ornamental species. With them, no doubt, were transported the early stages of numerous moths. Here, the microlepidoptera would have an advantage, being less noticeable and less easily dislodged, especially the leaf-miners, leaf-rollers, and those inhabiting spinnings. Thus many of the moths involved are associated with crops, ornamental shrubs, or fruit trees and bushes like currants, pears, plums and apples.

This explains how the moths arrived; the other question is why so many European species were able to establish themselves successfully, when North America had its own indigenous moths. I suggest this is because of the scale and rapidity of the habitat changes wrought by human settlers. Farming, forestry and urbanisation created an almost instant abundance of new niches. Given time, the native American moths would doubtless have managed to fill these niches: perhaps a few hundred years for existing species to adapt to new habitats and food plants, thousands of years (or much longer) for new species to evolve. This may still happen. Meanwhile, introduced moths have filled the vacuum. They were already adapted to human-modified habitats and to the imported plants, and found few native competitors.

The process has not yet finished. One of the most recent introductions is the Large Yellow Underwing, which arrived in Nova Scotia in 1979. By 1997 it had spread down the East Coast as far as the Carolinas and was breaching the Mid West (J Himmelman, personal communication). Perhaps the only surprise is that it took so long. It is easy to imagine a mated female of this abundant species flying aboard a passenger jet after dark at a British airport, and flying off again at the other end, adding to the 1,700 alien insects known to have become established in the USA by 1982 (Sailor, 1983).

Yet hardly any moths have come the other way. Even in Europe as a whole, only a handful of New World moths have become established. The American Wainscot (typically, called the Armyworm Moth in its home country) has colonised southern Europe, but is only a scarce migrant to Britain. To date, the only North American species to have colonised Britain are three micros (Appendix 3), the latest being *Argyresthia cupressella*, found in Suffolk in 1997 (Agassiz and Tuck, 1999). This is not because adventives never reach here: there are many records of American species from ports and docks, and from imported produce, while those detected are doubtless only a fraction of the true number. Plenty of American plants are widely grown in gardens, and vast areas of Britain are now planted with Sitka Spruce, Lodgepole Pine, and Douglas Fir, but none of their many associated macro-moths has accompanied them. Perhaps, in time, some will, and add interest to our very species-poor commercial forestry plantations.

Chapter 4

Identification

For the beginner, nothing is more frustrating than being unable to identify the species under study, whether it is a plant, bird or moth. For the expert, nothing is more exciting – it could be something new, or at least very rare. Getting from one stage to the other is a slow process, but anyone can do it, and there is no time limit.

The first step is to reduce the task to more manageable proportions. For various reasons, it is best for those beginning to study moths to concentrate on the macros at first. Taken as a whole, these moths are easier to identify by wing pattern alone. Except for a handful of species, it is possible to identify a live moth without the need to collect it as a specimen, or if in doubt, to take a reasonably competent photograph and ask a more experienced lepidopterist for help. Not so with the micros. These include several large genera of tiny and very similar species that can only be identified with certainty as adults by dissection under a microscope. Anyone who feels uneasy about killing the moths being studied is at an obvious disadvantage here. Most of the micros are too small to photograph without specialist equipment. Others, for instance the 60 or so British species in the genus *Stigmella*, are more easily recorded in their early stages by the characteristic mines the caterpillars make in the leaves of their respective food plants. Sometimes it is the caddis-like case that the caterpillar constructs for itself which provides the best clue, as in the genus *Coleophora*, with just over 100 species known in Britain – and approaching 400 in Europe. These really are groups for the specialist! The beginner should feel pleased simply to be able to allocate such moths to the correct genus, or even family.

Nevertheless, those who do begin to study the microlepidoptera often become totally captivated by them, forgetting most of what they once knew about the larger moths. Perhaps it is a human trait to be fascinated by miniaturisation. Unfortunately, as yet there is no popular identification guide covering all the microlepidoptera. Much of the literature for some families is scattered, and hard to obtain.

Guidebooks

In contrast, some excellent identification guides are available for macro-moths. The best of these is Bernard Skinner's *Colour Identification Guide to Moths of the British Isles*, revised in 1998, which includes every species you are ever likely to see in Britain (and dozens that you probably never will). Before then, the two volumes of *Moths of the British Isles* by Richard South (1907–09) held sway for over 70 years. So many were sold that second-hand copies are numerous and fairly cheap. It is still a book well worth having providing the copy is from the first edition, with good photographic plates (the quality varies), rather than the later versions where the illustrations are drawn by hand. As already mentioned, the English names used by South are almost unchanged today, though not the scientific ones, nor the order of classification.

Both these guides operate mainly on the principle of visual recognition – neither has keys. (I never find keys easy to use anyway, not having the right sort of tidy, logical mind.) Besides, for moths, keys are of limited value. Similar species differ not so much by firm, definitive structural features, as by subtle combinations of wing shape, colour and pattern. Furthermore, many moths are variable. Anyway keys seem to be designed for dead museum specimens (and curators): are the moth's eyes hairy, glabrous, or lashed? Is vein 5 on the hindwing strong, and connate with vein 4 at its base? Try doing that with a live moth, dashing around its container! Perhaps I am incompetent, but whenever I carefully check out a moth whose identity is already known, just to test a key, more often than not it gives me the wrong answer.

Live moths look different

The most straightforward way to identify an unfamiliar moth is to scan through the plates of a photographic guide to try to recognise it. Many beginners find there is a problem here: so many of the moths look the same. This is partly because the guides show moths in the spread-eagled position of set specimens, revealing the hindwings. The natural resting posture is usually totally different. Then, the hindwings are often concealed. Wings, body, legs and antennae may be held in a characteristic way, providing useful clues. Patterns that in the live moth run together across both forewings,

Figure 4: Three stages of wear (from top: newly emerged; about 7–10 days old; moribund) of a typical noctuid.

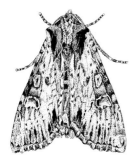

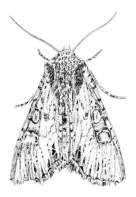

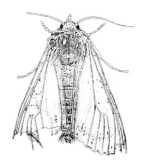

and sometimes across the body too, are disrupted in the set specimen. In my early collecting days, I often had to set a moth before I could pick it out in the guide. It is not as if the hindwings are of much use for identification anyway in most species, excepting the various 'underwings'. Disembodied hindwings from moths predated by birds or bats are usually impossible to identify with certainty if the forewings cannot be found.

Today, most people will not want to kill and set their moths. Experience makes it easier to visualise what a set specimen would look like if alive, and vice versa, by analogy with related species employing similar habits when at rest. But there are other hidden pitfalls when using a field guide. Naturally enough, the moths chosen to represent their species tend to be attractive, well-marked individuals, clearly showing all the diagnostic features. Not all the examples you catch will look like that: the colours may be duller and the pattern more obscure. Furthermore, the illustrated specimens are normally in pristine condition. Many will have been bred in captivity, and never flown. The moths you find in the wild, on the other hand, will show every gradation from perfectly fresh to badly worn and faded. Colours bleach, markings become less sharp, and in a really worn moth may almost disappear altogether. In this state, a moth can be impossible to identify by outward appearance alone (though examining the genitalia under a microscope would still give the answer).

As a boy, I was puzzled for years by a moth that my schoolmates regularly described to me. Its main feature was a red head; the wings were almost transparent, and often had a jagged outline. Somehow, I never came across this moth myself, nor could I find its like in the books. Many years later, I realised that they were describing almost any very worn noctuid. The 'red head' was a bald patch on the thorax, where the hair scales had been rubbed away, exposing the chitin. The wings were transparent because their scales had likewise been lost, and their edges were ragged not dentate (Figure 4). It is hardly worth wasting time trying to identify a moth in this condition, unless there is some pressing reason to do so.

Recognising variable species

Some moths have several distinctly different forms (polymorphism), in others the variation is continuous, one form grading smoothly into another. Frequently both sorts of variation are combined. However big the catch, it may be hard to find two Clouded Drabs or Lunar Underwings *Omphaloscelis lunosa* (Plate 54) that are alike, while the Dark Marbled Carpet (Plate 35) and July Highflyer (Plate 36) have already been discussed. No guidebook can illustrate every form. While even the expert can be fooled now and again, such infinitely variable species soon become easy to recognise.

Plate 54: Some moths are extremely variable, yet they can still be recognised by their 'jizz': the combination of shape, build, colours and pattern. These are three forms of the Lunar Underwing *Omphaloscelis lunosa*, an abundant autumn species over most of Britain but with a relatively restricted distribution in Europe. (Dorset).

This confirms that the birder's concept of 'jizz' applies equally well here. Just as a budgerigar or a feral pigeon is instantly identifiable whatever its colour and markings, so with variable moths. Somehow, the human brain is able to pick out the salient features of size, shape, structure, build, and the basic underlying wing pattern, all of which are relatively constant for each species.

Most variation in moths follows a few basic rules. It might be likened to the adjustments that can be made to the test card on a television screen. This can be made much lighter or darker. Contrast can be increased or decreased. The colour balance can be changed. Whatever permutations are made, it is still the same basic picture. Likewise with the markings on a moth's wings. Variation can make cross-lines heavy, bold and conspicuous, or so faint that they are almost obsolete. However, the shape of the cross-line stays the same: it will not change from being sharply angled and become gently curved. Spots and blotches may grow larger or smaller, merge or disappear – but only within the framework of the underlying pattern. Recognising that pattern enables the moth to be identified – though it isn't always as easy as that! The most extreme variations, often known as aberrations, do not necessarily follow these rules. However, they are very rare.

Weighing up the evidence

Sometimes it will be simple to pick out the moth to be identified in the guide, if it is a distinctive species and there are no others like it. Those are the easy ones. The numerous brindles, brocades and rustics are a different matter. Some of the carpets and nearly all of the pugs can be difficult too. The moth in question might resemble any of half a dozen species in the book.

Figure 5: Wing markings and structural features commonly referred to in descriptions of moths. (Above: a typical noctuid; below: a typical geometrid.)

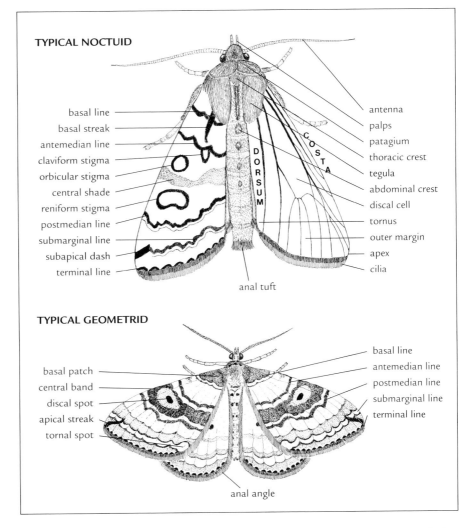

TYPICAL NOCTUID

basal line
basal streak
antemedian line
claviform stigma
orbicular stigma
central shade
reniform stigma
postmedian line
submarginal line
subapical dash
terminal line

anal tuft

COSTA
DORSUM

antenna
palps
patagium
thoracic crest
tegula
abdominal crest
discal cell
tornus
outer margin
apex
cilia

TYPICAL GEOMETRID

basal patch
central band
discal spot
apical streak
tornal spot

basal line
antemedian line
postmedian line
submarginal line
terminal line

anal angle

It might even look like a cross between two of them. This tempting fudge can be ruled out, as moths rarely hybridise under natural conditions. Crosses between some related species are possible in the laboratory, but I have never knowingly seen a hybrid moth or butterfly in the wild. So the lepidopterist does not have the complications that the botanist must face, where hybrids can be commoner than either parent, as with the marsh and spotted orchids.

That leaves a straight choice between several possibilities. Make a shortlist. Try looking at illustrations of those species in several different guides, if available. Features not so obvious in one sometimes show up more clearly in another. For confusing pairs of species, the text will stress distinguishing points. Figure 5 illustrates the main wing markings of a generalised noctuid and geometrid, and the terms used to describe them.

After selecting likely candidates on appearance, then consider the circumstantial evidence. First, the flight time of the moth: if the book says a moth flies in June and July, it is most unlikely that one caught in April is that species. In an early year, it might be out from the middle of May, especially in southern England; in a poor summer, it might linger on into August, especially in the north. But if the date is wildly wrong, the odds are that it is not that species. Next, consider distribution. For instance, if the books say the moth is found only in Scotland, and yours was caught elsewhere, virtually rule it out. Distribution maps help here. For all macro-moths except the Geometridae, the appropriate volume of *Moths and Butterflies of Great Britain and Ireland* gives a map for each resident species and the commoner migrants. If there are no 'dots' for a species anywhere near your area, don't expect to find it, unless you live in a very under-worked part of the country. Many English and Welsh counties have now published check-lists or atlases for Lepidoptera, and these are helpful. Then consider whether the habitat is right. Is the food plant present, and in sufficient quantity? Moths are mobile of course, and stray individuals can turn up almost anywhere. Even so, the best chance of finding a moth is in its correct habitat. Finally, does the guidebook describe the moth as common or rare? If it comes down to a choice between an abundant, polyphagous species found in a very wide range of habitats throughout Britain, and one that is scarce and has a restricted distribution, the strong probability is that you have found the former rather than the latter. Mistaking a common species for a rarity is embarrassing, and messes up the records if your identification is accepted (Plate 55). Far better to err on the side of caution, and be pleasantly surprised if a moth turns out to be a rarer species than you first thought.

Plate 55: Moths of the genus *Shargacucullia* are very similar as adults. As a result, dubious British records of several otherwise Continental species befog the literature. Most were probably misidentifications of the Mullein Shark *S. verbasci* (shown here), a common resident, observers having been too eager to claim it as a rarer species. Note how the warningly coloured caterpillar (Plate 90) gives rise to a cryptic adult. (Sussex).

Seeking expert help

Inevitably, there will still be moths you can't identify at all, and others where you cannot be absolutely certain of having done so correctly. This is frustrating, but there are several possible responses. The simplest is just to accept that you won't be able to identify every moth at first. Be content with recognising just some of the moths you find: the proportion will inevitably rise as you become more experienced. Congratulate yourself on those you do know, rather than agonise about those you do not. Moths are there to be enjoyed.

However, perhaps you are one of those obsessive personalities who are unable to take such a laid-back approach. Fine: many of the best moth people are! Faced with an unfamiliar species, you could collect it, set it if you wish to form a collection, and show it to an expert when the opportunity arises. Alternatively, a reasonably sharp photograph would do just as well for the great majority of macro-moths, and saves killing them, if that is an issue for you. Or perhaps there is a local lepidopterist who would be happy to help with identification – especially if might be something unusual or interesting. Most moths can be kept for a day or so in a cool refrigerator without harm, if not allowed to freeze or dehydrate.

Learning to recognise families

At the beginning, it may be necessary to look through all the plates in the guide to locate a moth, but soon it will become easier to recognise the probable family, and go straight to the right section. Even many subfamilies are distinctive. Knowing nothing about the moths of the USA, I could not name those found while on holiday there, yet could easily place most of them in the correct subfamily. The Noctuidae and the Geometridae are by far the two largest groups of macro-moths. Between them, they comprise 80 per cent of the species in Britain. In light trap catches, the proportion is often even higher, because many of them are very numerous, and the noctuids in particular come readily to light. Most of the smaller families are distinctive too, sharing features that soon become recognisable. Useful characteristics common to most, but not necessarily all, members of the main macro families (those with five or more species in Britain) and some important subfamilies are given below.

Swift Moths: Hepialidae (5 species; Plate 1).

Caddis-fly shape and resting posture. Antennae very short, as if broken off. Mainly crepuscular. No proboscis, so never found at flowers or sugar.

Burnets and Foresters: Zygaenidae (10 species; Plates 61, 70, 72, 124).

Foresters metallic green but inconspicuous. Burnets unmistakable: dark green with crimson spots. Both groups colonial and entirely diurnal, often nectaring at flowers.

Clearwings: Sesiidae (15 species; Plates 26, 98).

Wasp-like, but wings have dark fringes (cilia), borders and discal spot. Day-flying, but not often seen, except when newly emerged.

Eggars: Lasiocampidae (11 species; Plates 67, 73).

Stout-bodied, hairy, often large. Eyes relatively small, legs short. Wings usually broad and rounded with simple pattern of two cross-lines. Strong, feathered antennae in males. No proboscis, so never seen at flowers or sugar, but males diurnal in two species.

Hook-tips: Drepanidae (7 species; Plate 17).

Smallish, lightly built moths. Wings usually falcate (hooked) at apex. No proboscis, so never feed. Males diurnal in two species.

Lutestrings: Thyatiridae (9 species; Plates 81, 110).

Medium build and size, resembling noctuids but not closely related. Forewings usually with many wavy cross-lines, with small and cramped stigmata. Nocturnal; able to feed, frequently coming to sugar.

Geometrids or Geometers: Geometridae (295 species).

Emeralds: Geometrinae (10 species; Plate 106).
Both fore- and hindwings usually bright green (but soon fade), with paler or darker cross-lines. Nocturnal, often attracted to light.

Waves: Sterrhinae (38 species; Plates 24, 62, 125).

Small and delicate. Head fairly broad, with palps short. Usually bone-coloured, cream or brown with fine, wavy cross-lines. Hindwings coloured and marked like forewings. Mainly nocturnal, sometimes seen at flowers.

Carpets and Pugs: Larentiinae (162 species).

Carpets small to medium-sized, lightly built. Forewings basically triangular, often with an intricate, banded pattern. Hindwings usually plainer, and concealed at rest. Mainly nocturnal, but easily disturbed by day. (Plates 21, 28, 35, 36, 66, 79, 84, 119, 127, 129).

 Pugs very small, often brown or grey and obscurely marked, but resting posture of group is distinctive. (Plates 30, 48, 57, 96).

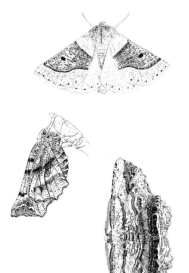

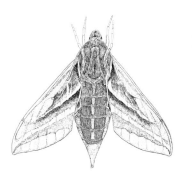

Thorns and Beauties: Ennominae (83 species; Plates 13, 15, 18, 19, 23, 29, 32, 51, 58, 63, 64, 68, 107, 108, 130).

Immensely variable in size, build (sometimes thickset), wing shape (often angled or scalloped), and pattern (often less complex than carpet moths). Resting posture is also varied, with hindwings often exposed at rest: if so, patterned like forewings. Females flightless in late autumn to early spring species. Strongly attracted to mercury vapour light, but not often to flowers or sugar.

A handy tip for recognising geometrids is that many hold their wings closed together over their back as do butterflies. For a few, such as Netted Mountain Moth, this is their normal resting position. Most of the carpets, and some of the Ennominae, also adopt it temporarily, when they are active at night or have been disturbed in the daytime. Fortunately, the waves and pugs never do, otherwise identification of live examples would be even harder. All moths adopt the 'butterfly position' when drying their wings, but as far as I am aware, only butterflies and geometrids ever use it again once they have flown. At least, this applies to British species, but elsewhere in Europe the Tau Emperor *Aglia tau* is an exception.

Hawk-moths: Sphingidae (18 species; Plates 2, 16, 27, 50, 74, 114, 122, 131).

Often huge and powerfully built, several being long-distance migrants. Forewings long, narrow and usually pointed; hindwings small. Some have a long proboscis and hover to feed at flowers, others cannot feed at all. Nocturnal species are strongly attracted to mv light.

Prominents: Notodontidae (27 species; Plates 20, 83, 85, 97, 112).

Trailing edge of forewing (dorsum) often with a triangular tuft of long scales forming a peak when the wings are closed. Legs furry, but all rather short. Most species attractively marked and distinctive. Nocturnal, unable to feed, rarely found at rest, but strongly attracted to mv light.

Tussocks: Lymantriidae (11 species; Plates 3, 123).

Moderate build, wings often broad and rounded. Female's abdomen often has large anal tuft. Forelegs usually long and furry, held extended when at rest. Males have very feathery antennae. Never feed, but come well to light.

Tiger Moths and Footman Moths: Arctiidae (32 species)

Footman Moths: Lithosiinae (17 species; Plates 7, 47, 103).
Slim build, many with long narrow forewings but broad hindwings. Rather drab, wing pattern sparse or absent. Many feed at flowers.

Tiger Moths: Arctiinae (15 species; Plates 25, 109).
Mostly stoutly built; warningly coloured and often boldly patterned, mainly with white, black, red or yellow. Lack proboscis, so never feed. Several are diurnal, but nocturnal species strongly attracted to mv light.

Nolids: Nolidae (5 species; Plate 128).

Small, white or grey moths, with raised tufts of scales on the forewings.

Noctuids: Noctuidae (397 species).

A vast and diverse group, its classification is currently in a state of flux. Bradley (1998) divides the British species into 19 subfamilies, as do Karsholt and Razowski (1996), but otherwise the two arrangements are very different, and further changes are inevitable. Nevertheless, the great majority of noctuids are easy enough to recognise. Most are stocky, medium-sized, nocturnal moths. Nearly all feed at flowers, many also at sugar. At rest, the forewings are either held roof-like, covering the hindwings, or overlap each other partially or completely. The forewing pattern generally includes a central kidney-shaped marking, the reniform stigma. Beside it, nearer the base, is the round or oval orbicular stigma. Often, the stigmata are a conspicuous feature. The lutestrings are the only other macros to have both these marks well defined (as do the Pyralidae among the micros, but these are much more slightly built). Some of the main groups of noctuids can be recognised as follows.

Darts and Clays: Noctuinae (60 species; Plates 34, 43, 71, 126).
Forewings often long and narrow, rather square-ended; hindwings ample. Stigmata usually obvious. Moth folds up like a penknife when at rest, forewings overlapping each other almost completely in some species. Legs long, especially hind pair.

Brocades, Brindles, Rustics, etc.: Hadeninae and others (many species; Plates 10, 22, 49, 54, 56, 65, 69, 75, 77, 104).
At rest, wings held in shallow roof-like posture (tectiform) with slight overlap, or not quite meeting, so exposing the abdominal crests where these are present. Wing pattern includes reniform and orbicular stigma.

Wainscots: Hadeninae and others (c. 30 species; Plate 33).
A type of wing pattern rather than a closely related group, evolved by many species associated with reeds and grasses. Forewings pale brown or whitish, streaked light and darker, giving a wood-grain appearance. Stigmata and other markings obsolete or absent.

Daggers: Acronictinae (14 species).
Usually some shade of grey, but never brown; pattern often includes black dagger-like marks. Moths rest exposed on tree trunks, walls or fences, so are often found by day.

Plusias: Plusiinae (22 species; Plates 46, 56, 111, 120).
Forewings have metallic silver or gold patches or marks, but stigmata are not obvious. Wings held in steeply sloping roof-shape at rest; crests on thorax and abdomen large and conspicuous.

Snouts and Fan-foots: Hypeninae and Herminiinae (15 species).
More slender than most noctuids, with a very triangular overall shape (thus also known as deltoids). On the head, elongated palps form a projecting, sometimes curved, 'snout'.

Identification by genitalia

So far, we have assumed that the moth under consideration can be identified by its outward appearance alone. However, a few pairs or groups of very similar species defy even the expert. These are often known as the 'genitalia species', because the only way to identify every individual with certainty is by examining the genitalia under a microscope or good hand lens. Sometimes it is necessary to prepare a slide first. Such specialised procedures are outside the scope of this book, but fortunately very few macro-moths fall into this category. Even among those that do, with practice it is usually possible to identify particularly characteristic individuals without having to dissect them.

The four *Amphipoea* species (Plate 56) are one such notorious group, especially as several may occur together. The Ear *A. oculea*, Large Ear *A. lucens* and Crinan Ear are all found on my own site in Banffshire. Eventually I learnt to identify about 70 per cent of moths from the group with confidence, and the rest with varying degrees of probability, simply by outward appearance. A few have to be left undetermined, but for general recording purposes this does not matter. Some of the clues I use might not work so well at other sites, but that is one advantage of really getting to know the moths in one's home patch.

Fortunately, it is not always necessary to kill moths oneself in order to examine their genitalia. I was fairly sure about some Pale November Moths *Epirrita christyi* in a Sussex beechwood, but two very similar species in the same genus are also found in that county. Checking the genitalia before submitting the record was essential. A visit to the wood following a mild but squally night solved the dilemma: several of the moths had drowned in puddles on the track, and the identification was soon confirmed. Likewise, to record the presence of both Common Rustic and Lesser Common Rustic *Mesapamea didyma*, it is easy to find traffic casualties (before the birds do) by the side of a busy road with grassy verges.

The largest of the difficult groups are the pug moths in the genus *Eupithecia*. They are the warblers of the moth world. A few of the 46 species listed for Britain are distinctively marked, but the rest are 'small brown jobs' that may confuse expert and beginner alike (Plate 57). Apart from melanic, worn or poorly marked forms of a few species, all can be identified without examining the genitalia – though sometimes it is easier to do that. Many observers are satisfied just to recognise the genus, without determining the species. For others, these moths are a challenge to relish, a real test of identification skills. It does not help that, being so small, their markings do not show up to best advantage in the guides. Picking out the right species from rows and rows of similar ones can

Plate 56: The four British *Amphipoea* species can be impossible to separate without dissection of their genitalia. However, the more characteristic individuals can be identified from their appearance, as follows:

The Ear *A. oculea* (top left). Small, short-winged; forewing dark, crosslines not prominent; reniform stigma usually white, rounded, with clear internal divisions. Throughout Britain. (Banffshire).

Crinan Ear *A. crinanensis* (bottom left). Medium-sized, broad-winged; forewing a clean mid-brown, crosslines and veins sharp as if finely pencilled; reniform stigma squarish, usually orange-yellow. The last to emerge; mainly northern and inland. (Banffshire).

Saltern Ear *A. fucosa* (top right). Wings longer and narrower; forewing often pale, sandy, crosslines usually faint; reniform stigma, if white, often reduced to a narrow crescent. Coastal, and mainly southern. (Suffolk).

Large Ear *A. lucens* (bottom right). Large and robust, with long wings; forewing richly coloured but appears sullied, often with darker central area, crosslines strong; reniform stigma somewhat narrowed, especially if white. Mainly northern and inland. (Banffshire).

seem a daunting task. Fortunately, circumstantial evidence is particularly useful in this group. Each pug tends to be restricted to a certain food plant, as reflected in many of the English names. No doubt this is how the numerous species diverged in the first place. Usually they do not stray far from their associated plant, so that a pug fluttering around the garden currants may well be the Currant Pug *Eupithecia assimilata*, one disturbed from the trunk of a Larch tree is likely to be a Larch Pug, and so on. At least, it makes sense to consider the eponymous species first. The loss of such helpful clues is of course one of the penalties for using a light trap instead of doing proper fieldwork!

Making contact with others

It is perfectly possible to become a solitary, self-taught expert on moths, learning everything from the literature and through personal experience. Far better, though, to make contact with others who share the same passion. Moths will always be a minority interest, but this means that new recruits are generally welcomed, and seen as potential friends and colleagues rather than rivals. There are still plenty of vacant niches! The over-competitiveness that sometimes afflicts birders is lacking amongst lepidopterists. Tell a twitcher that you missed seeing a certain rarity, and he will secretly exult at having 'gripped you off'. But tell a moth person that you have never seen a particular species, and nothing would give them greater pleasure than to take you to the best available site, in the hope that you will see it too. All true moth enthusiasts will gladly help with identification problems, engage in good-humoured debate, and pass on useful tips – or the kind of local knowledge that cannot be gained from books.

Today, it is easier than in the past to meet other people interested in moths. Increasingly, county wildlife trusts, and natural history societies such as Butterfly Conservation, hold Moth Nights which are open to all, when mercury vapour light traps are run, with an expert on hand to identify the catch. Having moths identified for you in this way can be useful, especially if the expert possesses teaching skill, but it is no substitute for doing it yourself. The social side is just as valuable, enabling contact with people with similar interests. Residential courses on moths are held at various field centres, often in very scenic parts of the country that are good for other wildlife. There are telephone hotlines, and internet sites too – so why is it still so fiendishly difficult to identify that little brown pug?

Plate 57: Identifying a grey, obscurely marked pug such as this is a tricky test for even the most experienced observer. It might be dismissed as one of the commoner species, but would merit a second glance if found in lime *Tilia* woodland, being Fletcher's Pug *Eupithecia egenaria*, a recent colonist. (Norfolk).

Chapter 5

Finding moths by day

Butterflies fly in the daytime, but moths are nocturnal. Most people with any interest in wildlife know that this is only a truism, and could name moths that are exceptions. (Migratory butterflies like the Red Admiral *Vanessa atalanta* are occasionally active at night too, and are caught in light traps.) In fact, the number of moths that do fly naturally in the daytime is surprisingly large. The information is not available for all species, but Emmet (1992) gives flight times of 1,063 British microlepidoptera (excluding the 'honorary macros'). Extracting this data shows that only 269, or just over 25 per cent, are regarded as wholly nocturnal. They are outnumbered by the 374 species (35 per cent) that are at least partly diurnal, though some of these also fly at dusk or at night too. The remaining 420 species (40 per cent) are crepuscular, flying in the evening, and sometimes also at night.

Among diurnal micros, the longhorn moths in the subfamily Adelinae are sure to attract attention because of the wholly disproportionate length of their antennae, up to four times the length of the wings in *Nemophora cupriacella* males. Why this is so is unknown. The bright little pyralid moths in the genus *Pyrausta*, forever associated in my mind with long hot days and the smell of wild thyme trodden underfoot, are also unlikely to go unnoticed. *Anthophila fabriciana* can be disturbed from almost any patch of nettles throughout Britain, and for that reason is sometimes known as the Nettletap Moth. The tiny Cocksfoot Moth *Glyphipterix simpliciella* is also found everywhere. It makes up in numbers what it lacks in size. I have seen eight inside one buttercup flower, and there was still plenty of room.

Macro-moths are usually nocturnal or crepuscular. Even so, about an eighth of our residents and common immigrants fly naturally at some time during the day on a regular basis. Appendix 4 lists 105 such species. Of these, about half can be regarded as largely or wholly diurnal, at least in one sex,

Plate 58: Latticed Heath *Chiasmia clathrata*. Unusually, this butterfly-like geometrid is equally active both in bright sunshine and at night, when large numbers sometimes appear at light traps. (Norfolk).

Plate 59: Like other members of its subfamily, the Gold Spot *Plusia festucae* normally flies from dusk, but occasionally nectars by day in hot weather. The purpose of its metallic spangles on a tawny ground is unclear; many fritillary butterflies have a similar pattern on the underside of their hindwings. (Banffshire).

nearly always the male. As with butterflies, sunshine is usually needed to make them active. The rest fly at certain times of the day, most often in the afternoon, then again at dusk or at night. However, the Purple-bordered Gold *Idaea muricata* famously flies at sunrise, a time when few Lepidoptera are on the wing – or maybe few observers are there to see them.

Sometimes there are regional differences in behaviour. In my part of north-east Scotland, the Double-striped Pug *Gymnoscelis rufifasciata* seems to be completely diurnal. It flies over the heather in the mid-afternoon sunshine, and in 10 years I have never seen it at artifical light. Yet in Sussex I never saw it flying before dusk, and often caught it at light. Likewise, the Ruby Tiger *Phragmatobia fuliginosa* becomes almost entirely diurnal in the north, perhaps because night temperatures are lower there.

Many moths other than the ones listed are occasionally seen flying in the daytime. In exceptionally hot and dry weather, some individuals are faced with the choice of either dying through dehydration, or becoming active at the 'wrong' time to find nectar sources (Plate 59). Even purely nocturnal species such as the Large Yellow Underwing are then sometimes seen feeding at flowers. Generally, they are rather worn individuals nearing the end of their life anyway. Recently arrived migrants greedily feed by day too, desperate to replenish their resources. The Silver Y, usually crepuscular, is a familiar example.

Numerous moths normally fly after dark, but readily take wing if disturbed in the daytime. Generally, these are the more lightly built species, with a low wing loading. The stockier moths, like prominents and most noctuids, need to warm up their wing muscles before they can fly. They do this by 'revving up' – vibrating the wings rapidly and with increasing vigour until the thoracic muscles reach the right temperature. Only if it is a really hot day (approaching 30°C) can these moths take off instantly. Otherwise, they usually react to threats by dropping to the ground and 'playing dead'. Most geometrids on the other hand, especially the carpets, can fly without any preparation. Some do so at the slightest disturbance – a shadow or a foot-fall – if the day is reasonably warm. The Grey Mountain Carpet and the July Highflyer are particularly skittish, even though each is beautifully camou-flaged against its favourite resting background, rocks and tree trunks respec-tively.

It follows that a good way to search for such moths in the daytime is to disturb them from their hiding places. Simply walking through vegetation such as long grass, bracken or heather will cause many species to fly up. They can then be netted, or followed until they settle again. This works best on warm, dry and calm days. If it is breezy, however, many will quickly be whisked out of reach. Gently shaking overhanging branches, tapping bush-es with a stick, and rustling tall vegetation will also cause moths to take wing. Shaded and sheltered places, such as the lee side of a wood or a sunken track, are the most productive.

Moths that are much less ready to fly in the daytime must be sought by eye. For obvious reasons, the many species that hide amongst leaf litter, grass tussocks or low vegetation are seldom found, except by a lucky chance. Others trust in their camouflage, and sit openly on trees and rocks, or those man-made equivalents, fences and walls. The shaded side is always pre-ferred, and moths will gradually move round a trunk or fence post to avoid strong sunlight. Gusty winds will also cause them to seek a more sheltered position. In the past, searching for these species used to be a major activity of the moth hunter, but such is the efficiency of modern light traps, it is often considered unnecessary today.

There is no doubt that daytime searching for moths that can be caught much more easily by other methods is not cost-effective in purely ergonom-ic terms. Rather, it should be looked upon as a measure of skill, eyesight and experience. Finding just one moth at rest in its natural surroundings can give far more pleasure than catching a dozen of the same species at light. It also gives the photographer a chance to obtain that genuinely unposed, taken-in-the-wild picture – much more satisfying than any rigged-up studio shot. Also, daytime searching can be used at sites where light trapping is not prac-tical. Access might be restricted at night, the terrain potentially dangerous, or the chosen spot too far to carry a trap plus its heavy generator or bat-tery. Searching for moths requires no equipment! But best of all, in daylight the habitat too can be seen and enjoyed, together with its other wildlife. Scenery and views, trees, flowers, fungi, birds, butterflies, dragonflies all help to make an expedition enjoyable, whether any interesting moths are found or not. The day I saw my first Broad-bordered White Underwings was equal-ly memorable for the pink mats of Trailing Azalea where some were nec-taring, and for the pair of Ptarmigan, more puzzled than alarmed by human trespass into their territory.

Downland

In general, habitats that are good for butterflies are also good for moths. However, the short flowery turf of chalk downland is not quite as outstanding for moths, especially after dark. With sparse vegetation cover, the convex slopes quickly radiate heat once the sun goes down. Even after the hottest day, nights can be very cold, and moths soon become inactive. Perhaps because of this, many of downland's characteristic species are day flying, like the foresters and burnets, *Pyrausta nigrata*, Mother Shipton *Callistege mi*, and the Small Purple-barred *Phytometra viridaria* (Plate 60).

Some of these diurnal moths are not as conspicuous as might be assumed. For over 20 years I worked a splendid piece of downland in Sussex, first as a collector, later as a voluntary warden once it became a nature reserve. Most of the expected chalk-loving butterflies and moths could be found there, often in abundance. There was a casual report in the office files that one of the forester moths had been seen, but with the cocksureness of (comparative) youth I dismissed the possibility. No such species could occur on my patch without my knowledge! Then, one sunny day, a visiting botanist remarked on the metallic green moths that were sitting around on Salad Burnet flowerheads. They were Scarce Foresters *Jordanita globulariae* (Plate 61). In flight, they were hard to see against the green turf, and did not look like moths at all. Their wings seemed almost transparent, so that they resembled sawflies, lacewings, small grasshoppers or others among the myriad of flying insects that were traversing the downland slopes. Sometimes it helps to have, like that botanist, no preconceptions about what is or is not supposed to be there, and what it should look like.

In very open downland, it is well worth investigating any isolated taller vegetation, such as a patch of brambles, tussocks of Tor-grass, or the clumps of nettles, Hound's-tongue and ragwort that often grow in the excavated earth of badger sets. Moths are often concentrated in these few patches of cover, and a dozen or more geometrids sometimes fly out when a clump is gently disturbed. Inevitably, most will be the Yellow Shell (Plate 84), but there is always the chance of something scarcer. The Dusky Sallow has a strange habit of sitting all day on knapweed heads, where it is extremely conspicuous against the purple flowers. Its colour and pattern suggest that the dry seedheads of Cock's-foot grass, the larval food plant,

Plate 60: The Small Purple-barred *Phytometra viridaria* is a good indicator of herb-rich short turf, whether downland or heathland. Highly active in sunshine, this one swivelled to face the camera, providing a welcome alternative to the standard upperside shot. (Banffshire).

would be the natural choice; no doubt those that do sit there are usually overlooked (Plate 49).

Other moths prefer bare chalk and scree, as found in quarries, where they rest on the exposed rock. These include the Chalk Carpet *Scotopteryx bipunctata* and Mullein Wave *Scopula marginepunctata*, which has evolved a special almost white form (Plate 62) on the Downs. In recent years, the cuttings, embankments and wide verges formed during the construction of major new roads have created prime habitat for these and other downland moths, so if access is permitted these are good places to look.

Today, few large pieces of open downland remain, following the breakdown of the traditional sheep-walk system that created them. Agricultural changes since 1945 have seen vast areas ploughed and sown to barley. Most of the remnants were invaded by pioneer scrub after grazing ceased, and myxomatosis decimated the rabbit population. This was devastating for butterflies such as the Silver-spotted Skipper *Hesperia comma* and Adonis Blue *Lysandra bellargus*, which depend on very short turf. For moths, the effects were mixed. Scrub, especially if varied, greatly increases the overall number of moths that downland can support. However, most of these additions are relatively common species, not confined to the chalk. Their gain does not compensate for the loss of the true downland specialists like the Lace Border *Scopula ornata* and Straw Belle *Aspitates gilvaria*, now both much scarcer, or the Feathered Ear *Pachetra sagittigera*, now thought to be extinct.

Nevertheless, some very attractive moths are found in downland scrub. Where there is Spindle, the Scorched

Plate 61: Scarce Forester *Jordanita globulariae*. Restricted to the best chalk downland, this is a very local moth, albeit easily overlooked when numbers in a colony are low. (Sussex).

Plate 62: Mullein Wave *Scopula marginepunctata*. About 10 per cent of some East Sussex colonies are of this form, the clear white ground colour concealing them at rest on bare chalk.

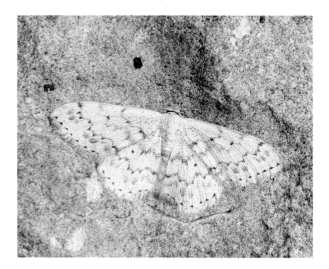

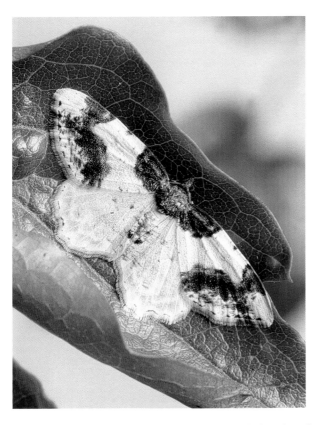

Plate 63: Scorched Carpet *Ligdia adustata*. Especially common in downland scrub, this moth rests openly. Perhaps birds either mistake it for a bird dropping, or recognise it as unpalatable by association. (Sussex).

Carpet sits openly on foliage, camouflaged as a bird dropping (Plate 63). Buckthorn supports a related group of three rather local geometrids, including the Tissue. Like the Brimstone *Gonepteryx rhamni*, our only butterfly associated with Buckthorn, it hibernates as an adult, which might not be coincidence. Older downland scrub is often festooned with skeins of Traveller's-joy, the food plant of another group of geometrids. One of these is Haworth's Pug *Eupithecia haworthiata*, so small and dull even for a pug that Doubleday was hardly paying that eminent early lepidopterist a compliment by naming the species after him in 1856. Haworth, though, could not object: he died in 1833. The others in the group are much more attractive, especially the Small Emerald *Hemiostola chrysoprasaria* before its delicate green tint fades, and the disruptively patterned Small Waved Umber *Horisme vitalbata*.

Chalk downland that has been turned to arable land supports very few moths, except for the narrow strips along headlands, tracks and fence lines. There is one way this can be used to advantage. In Sussex, I regularly used to walk through the barley stubble after the harvest. Once away from the edge of the field, it was unusual to disturb a moth, but if so there was a fair chance it would be a recently arrived immigrant. Silver Ys and the Rush Veneer *Nomophila noctuella* predominated, but it was a good way to find the Vestal if there were any around that year. No doubt there were just as many on the better downland, but there they were harder to pick out from the numerous resident moths of similar size and colour.

Woodland

Most scrub eventually turns into woodland. There is no doubt that ancient semi-natural deciduous woodland is the best habitat of all for moths in terms of the number of species it holds. Ideally, there should be a wide range of native trees, especially oak, birch, Ash, elm, Alder and Aspen. Some should be very old, with lichens and bracket fungi, others still saplings. The shrub layer should include sallow, wild rose and Hawthorn, and there must be wide rides and sheltered glades with a rich ground flora. Few present-day woods match this ideal, but even average woodland will suffice for all but the most demanding moths.

The caterpillars of many of our most characteristic woodland butterflies do not feed on trees. That of the Speckled Wood *Pararge aegeria* feeds on grasses, the Wood White *Leptidea sinapis* caterpillar on vetches, and those of the fritillaries on violets. The same is also true of woodland moths, if to a lesser extent. Many of their caterpillars do indeed feed on trees and shrubs, but for others it is the sheltered environment that matters, and the food plants are grasses or low plants that are by no means confined to woodland themselves. Somewhat surprisingly, no specialist moths compete with the fritillary butterflies for violet leaves, though doubtless these are eaten by a few polyphagous species. Emmet (1992) lists only two moths associated with violets, *Pancalia leuwenhoekella* and *P. latreillella*; both mine the rootstock and are grassland or heathland species anyway.

Plate 64: Speckled Yellow *Pseudopanthera macularia*. As usual, we can only speculate whether the colours of this diurnal woodland moth camouflage it in dappled sunlight or signal that it is unpalatable – or both. (Banffshire).

Although so many species are present in woodland, as revealed by light-trap catches, paradoxically it can be the most difficult habitat in which to find moths by day. Relatively few woodland macro-moths are diurnal. Perhaps there are so many different insectivorous birds waiting to prey on them. The microclimate at night is more reliable, compared with open habitats, because of the shelter and the heat-retaining effect of the leaf canopy, so perhaps there is less need to fly in the daytime. However, as always, some moths are exceptions. The clearwings are active in sunshine, but rarely seen once they have flown after emergence. More conspicuous are the brightly coloured Speckled Yellow *Pseudopanthera macularia* (Plate 64), and the little coal-black Chimney Sweeper (Plate 28). In both cases, the caterpillar feeds on a woodland herb rather than a tree, Wood Sage and Pignut respectively.

Crepuscular and nocturnal moths are often hard to find in mature woodland. Some species are well-concealed in the leaf litter, but many others are out of reach, resting in the canopy or on trunks, branches and twigs far above the ground. Even when the searcher does manage to disturb a moth that was resting lower down, all too often it flies tantalisingly higher and higher, evading capture, before settling again. In effect, the lepidopterist is restricted to the lowest couple of metres of a habitat that may be 12 metres tall or more, whereas birds can search it all. Watch mixed parties of tits and warblers in July or August, working their way through the treetops, and see how many moths they put to flight. Those are just the ones that get away.

However, some desirable moths can still be found by searching. A traditional way to obtain the Rannoch Sprawler is to examine the trunks of

Plate 65: Rannoch Sprawler *Brachionycha nubeculosa*. For nearly 150 years, English collectors have visited the Scottish Highlands in early spring for this rarity. Sadly, light trapping has largely superseded traditional searching. (Aviemore).

mature birch trees in its few Scottish haunts. The trunks are gnarled, with deep fissures and crevices, and a good covering of lichens. The moth is extremely well camouflaged. David Barbour (personal communication), who has found more than most (including the female shown in Plate 65), estimates that on average it takes 6 hours of intensive searching to find one moth. Fittingly, April 1st is held to be the most propitious day for this activity. Success is far from guaranteed, but striking northern forms of common spring birch-wood species like the Yellow Horned (Plate 110) and Early Tooth-striped *Trichopteryx carpinata* help to stop the concentration flagging, and the scenery is superb.

Coniferous woodland is generally poor for moths in Britain. Worst of all are forestry plantations, especially monocultures of non-native trees like Sitka Spruce and Douglas Fir. Not only do these trees have very few moths associated with them here (unlike in their homelands), but the artificially close spacing also prevents the growth of any shrub layer that might support additional species. Often, in the deep shade, there is little or no ground layer of vegetation either, except along rides and firebreaks. Larch is slightly better: although it has few specialists, a few caterpillars that normally feed on broad-leaved trees will eat it, like the Scalloped Hazel and the Engrailed *Ectropis bistortata*.

However, being one of our two native conifers, Scots Pine is comparatively good for moths (Plate 66). In Britain, it supports 32 species, three-quarters of which are micros, including many tortricids. Where it grows naturally, in the remnants of the ancient Caledonian forest, the trees are widely spaced and admit sunlight. This permits a dense ground layer of vegetation to form. Bilberry often predominates, and is excellent for moths in its own right. It is the food plant of the Rannoch Looper *Itame brunneata*, and is also eaten by caterpillars of a wide range of species that are polyphagous either on shrubs or on low plants. One is the Mottled Beauty, which rests as an adult, beautifully camouflaged, on the pine trunks. Even so, in the same part of the country, broad-leaved woodland such as birch will support several times as many moths as does Scots Pine.

Plate 66: Many of the moths associated with pine adopt the reddish tints that camouflage them against the bark of their host tree – well shown by this Pine Carpet *Thera firmata*. (Banffshire).

Heath, Moorland and Mountain

Lowland heath, what little of it has not been lost to agriculture or building, is an important habitat for moths. It cannot compete with woodland in terms of sheer numbers, but it supports an interesting community of species. Some of these depend entirely on heathers and Bilberry, others on gorse and Broom. Many, however, feed on vetches or other low plants as well as on heather, but favour the general heathland environment.

Once, heaths were common land, maintained as open ground by regular grazing, cutting and controlled burning that prevented the regeneration of trees. Today, most are no longer grazed, and are reverting to woodland, like much of the Ashdown Forest in the Sussex Weald. Birch, pine and sometimes oak saplings at first diversify and increase the number of moths present, but eventually cause the loss of both the heath and its heathland species even if they have survived the destructive forces mentioned earlier. The plight of the Smooth Snake, Sand Lizard, Woodlark and Dartford Warbler has long been documented; equivalent southern rarities among the moths are the Speckled Footman *Coscinia cribraria* and Shoulder-striped Clover *Heliothis maritima*.

Many of the characteristic heathland moths that are not tied to heather as a food plant are also found, less commonly, in other open habitats such as downland. A surprising number are day flying. They include the Latticed

Plate 67: Fox Moth *Macrothylacia rubi*. Males of this characteristic heathland moth are usually seen in their powerful, erratic late afternoon flight in search of females. This one was discovered warming up in readiness. (Banffshire).

Heath *Chiasmia clathrata* (Plate 58) and Common Heath in both sexes, and the Emperor (Plate 52), Fox Moth (Plate 67) and Oak Eggar, where only the male is diurnal. The moths in the latter group are large, and endowed with stores of fat because they do not feed as adults. They are more easily seen than caught, except by falcons.

Several of the Arctiidae are particularly associated with heaths, though not confined to them; for example the Wood Tiger (Plate 109), Clouded Buff and Ruby Tiger, all of which are at least partly diurnal, plus the Muslin. The latter is an oddity amongst British moths because it is the female that flies in the daytime, whereas the male is strictly nocturnal, a reversal of the usual roles. The Four-dotted Footman is easily put up from the heather by day, although night is the usual time of flight. Its caterpillar eats lichens growing on heather, not the heather itself.

In upland and northern Britain, heath becomes moorland, a bleaker and more extensive habitat dominated by dwarf shrubs. These are present in greater variety: Cowberry, Bog Bilberry, Bearberry, Crowberry and Bog-myrtle can often be added to the Ling, bell heathers and Bilberry found on heathland in the south. The moths also contradict the general rule of fewer species with increasing distance northwards: although some southern moths like the inappropriately named Horse Chestnut are absent, they are replaced by a much greater number of moorland specialists often loosely known as Arctic-Alpine species.

Typical of this group are the Netted Mountain Moth (Plate 68) and the Small Dark Yellow Underwing. Both are associated with Bearberry, especially where this grows on stony glacial moraines. Being relatively firm and well-drained, these moraines were often used in eighteenth-century Scotland to route military roads. No longer needed to quell the rebellious clans, the roads now provide easy access for the lazier lepidopterist who wishes to see these moths, in otherwise remote and difficult country. Fortunately for the purist, there are still sites for the species that can only be reached after a long, arduous trek!

Moorland is perhaps the easiest habitat of all to work for moths. Many of its species are diurnal, like the two just mentioned. Others are very easily disturbed in the daytime simply by walking through the heather, some of which occur abundantly, like the Northern Spinach *Eulithis populata* and Satyr Pug *Eupithecia satyrata callunaria*. Even the wholly nocturnal species tend to sit around fully exposed in the daytime on rocks and fence posts, where they

Plate 68: One of the Scottish 'specialities', the Netted Mountain Moth *Macaria carbonaria* is misnamed in that it frequents Bearberry moorland, not summits. Sunshine is necessary for the moths to fly. (East Inverness-shire).

Plate 69: Searching moorland fenceposts in late May is the best way to find the Glaucous Shears *Papestra birens* during its relatively short flight period. It also rests on rocks, where it is more difficult to spot. (Banffshire).

may be found by careful searching. Moorland fences are particularly productive, because in treeless areas nothing else stands up above knee height, giving moths that rest higher no other choice. The Rannoch Brindled Beauty (Plate 29), Glaucous Shears (Plate 69), Light Knot-grass *Acronicta menyanthidis* and Sweet Gale are among many that can be found in this way. It also helps that their camouflage does not work so well against the sawn wood of fence posts, unless these are very weathered and covered with lichens. The usual rule that moths prefer the shaded and sheltered side applies, except in the case of the Small Dark Yellow Underwing, where mating pairs bask in the full afternoon sun. However, not all fences are equally good: certain ones regularly harbour interesting moths, whereas others prove consistently disappointing for no obvious reason.

Many of the moorland moths have an altitude limit, being found most abundantly on the lower slopes and gradually petering out as the ground rises. There is a small but select group where the converse applies. They are mountain-top species, not usually being found below about 2,000 feet (800 metres) above sea level in mainland Britain. Only four of our macro-moths fall into this category, including the Broad-bordered White Underwing already mentioned. All have a certain charisma because of their lifestyle and the effort required to observe them in their rugged haunts.

Where they do occur, they are often the commonest, or even the only, macro-moths present. To those familiar with its relatives in downland or coastal localities, nothing could look more incongruous than a colony of the Mountain Burnet *Zygaena exulans* at one of its sites near Braemar, amongst stunted heather and Crowberry on a windswept summit plateau (Plate 70). In this harsh environment, with its short cool summers, the caterpillars take several years to reach full growth, depending on how many times they go into diapause.

The Black Mountain Moth has a 2-year life cycle, mostly spent as a caterpillar, as does the Northern Dart (Plate 71). Like the other mountain moths, their caterpillars feed on Crowberry. However, both these species show periodicity – moths are on the wing only every other year, odd and even numbered ones respectively. The reason for this is still not fully understood: what prevents the system from breaking down if some individuals take 1, or 3, years to complete their life cycle?

Parasitoids offer the most likely explanation. In June 1996, I found 11 pupae of the Northern Dart on a mountain-top in Banffshire. Only three went on to produce moths. A single large parasitoid wasp, *Ichneumon nigroscutellatus fennicola*, emerged from three others. Another had been 'stung', but nothing resulted. However, the four remaining pupae gave rise to no less

than 332 individuals (mostly females) of a tiny pteromalid wasp, *Coelopisthia caledonica*. This is probably the main culprit. So at this site in 1996, admittedly in a small sample, Northern Dart adults were outnumbered over 100 to 1 by their parasitoids.

Plate 70: Smaller and duller than its relatives, the Mountain Burnet *Zygaena exulans* is surprisingly hard to see amongst the short vegetation of its alpine haunts – here at over 3,000 feet above sea level. (Braemar).

Consider what would happen the following year. *C. caledonica* overwinters as an adult, so in late spring 1997 the females will be searching for suitable hosts. There will be plenty of half-grown Northern Dart caterpillars, but those are safe, as the female wasp lays eggs only in a newly formed pupa. With such a high population of wasps, any out-of-sequence 'wrong year' Northern Dart pupa is unlikely to escape. But most of the female wasps will fail to find a host. The parasitoid population will crash. There will be few wasps to attack the numerous Northern Dart pupae in summer 1998.

This perpetual leap-frogging of host and parasitoid populations is likely to result whenever a host with a 2-year life cycle is preyed on by a parasitoid with a 1-year cycle, in a relatively simple ecosystem where there are few if any alternative hosts or predators. It is hard to say whether the arrangement

Plate 71: Northern Dart *Xestia alpicola*. This charismatic mountain species is still most easily obtained by the traditional method of searching for pupae – but only in even-numbered years. The British subspecies, *alpina*, is much more handsome than the nominate race found in Fennoscandia. (Banffshire).

benefits the moth or the wasp more. Clearly both species can live with it, each having a guaranteed advantage every other year, which may be valuable in the harsh environment of a mountain summit. Equally, the lepidopterist knows in advance which years to go and look for the moth, and when not to bother.

Those who venture into remote areas after moths must of course follow the usual safety rules. Preferably, go with a companion; this is much more fun anyway. Wear proper boots and sensible clothing: the weather can deteriorate without warning up on the mountain-tops. Do not overestimate your fitness and capabilities. Take high-energy snacks, and a drink if there are no suitable streams. Do not carry unnecessary equipment – you may need to move quickly over difficult terrain. Let someone know where you intend to go, and when you expect to be back.

Coasts

In terms of quality rather than quantity, coasts are excellent for moths. Several factors combine to make this so. A few moths are found nowhere else because their larval food plant is strictly littoral, like Marram grass for the Shore Wainscot *Mythimna litoralis*, and the plant whose name it bears for the Lyme Grass *Chortodes elymi*. Many more species are less restricted by food plant, and also occur at some inland sites, but are predominantly coastal because of habitat preferences for sand or cliffs. Climate seems to be the key for other moths. Because the sea retains heat better than the land in winter, there are more frost-free days at the coast. For species at the northern edge of their range in Britain, the south coast of England may be the only place they can survive. The L-album Wainscot is exclusively coastal here, but not in the rest of its European range, and there are many similar examples. The same factor can be seen operating within Britain. The Six-spot Burnet and the Cinnabar, found throughout the southern half of England, become increasing coastal northwards and almost exclusively so in Scotland. Finally, as with birds, the coast is the best place to find newly arrived scarce migrants.

Unfortunately, daytime searching for moths can be unrewarding in coastal localities. In particular, many of the characteristic sand dune species are seldom found by day. Several belong to the subfamily Noctuinae, the darts, and moths in this group are notorious for hiding away very successfully in vegetation or debris at ground level. Although the Coast Dart, Archer's Dart and Sand Dart *Agrotis ripae* are often locally abundant, look-ing for them in the daytime is a thankless task. The books say that they can be shaken out of overhanging tufts of Marram, as can the Portland Moth, but my own attempts have never been rewarded. It is far easier to find such moths at night. This also applies to the various coastal wain-scots, concealed low in grass tussocks dur-ing the day, and protected by their cryptic wing pattern. Birds are more adept at the hunt: I once searched sand dunes for the Shore Wainscot, but found only wings in neat sets of four, discarded by Meadow Pipits.

Cliffs offer slightly more scope. Moths like the Annulet *Gnophos obscuratus* can, with difficulty, be detected at rest under overhanging ledges, or more often dis-turbed from scree. It is one of very few widespread geometrids (as opposed to rar-ities found only at a handful of sites) in Britain with a predominantly coastal dis-tribution. Another is the Rosy Wave *Scopula emutaria*, associated with salt marshes. Otherwise, in coastal habitats

Plate 72: The Transparent Burnet *Zygaena purpuralis*, though thinly scaled, is not so translucent as illustrations of set specimens suggest. In life, the red hindwings exactly reinforce the forewing streaks to give a strong colour. Many colonies are on dangerously steep and crumbling sea cliffs. (Argyll).

most of this family are general species of grassland, downs or moorland, though like the ubiquitous Yellow Shell and Shaded Broad-bar *Scotopteryx chenopodiata* they are often particularly abundant. In contrast, there are at least twenty basically coastal noctuids.

In western Scotland, sea cliffs are the habitat of several of the scarcer burnet moths, partly because more accessible areas are badly overgrazed by sheep and Red Deer. In general, an interest in moths is not a dangerous hobby, but some colonies of the Transparent Burnet *Zygaena purpuralis* (Plate 72) are on extremely hazardous and unstable slopes, where one lost foothold could trigger a slide that ends with a sheer drop into the sea. Nor is it wise to work directly below a companion who may be dislodging boulders higher up the slope.

Wetlands

Fens, marshes, wet meadows and the margins of water bodies have their own characteristic moths. Many are wainscots, associated with Common Reed and other tall waterside grasses, bulrush and Yellow Flag. The adults of this group are rarely encountered by day, as already mentioned, but some are readily found as caterpillars or pupae. Other wetland noctuids include the Small Clouded Brindle *Apamea unanimis*, Double Lobed *A. ophiogramma* and Gold Spot *Plusia festucae* (Plate 59), all associated with Reed Canary-grass. The latter is sometimes found resting on low vegetation, as is the Drinker Moth (Plate 73).

Only a few wetland macro-moths are diurnal. Most spectacular, but very localised, is the colonial Scarlet Tiger. Equally local and striking (once recognised as the mimic not the model) is the Narrow-bordered Bee Hawk-moth, as it darts and hovers to feed at flowers. The Ear moth is occasionally active in sunshine, and the more northern Crinan Ear regularly flies in the daytime, nectaring at Devil's-bit Scabious. But perhaps the most noticeable wetland moths are micros. The china-marks, large, long-legged and attractive members of the family Pyralidae, are often disturbed into lazy flight from waterside vegetation during the day, although they are mainly crepuscular. The numerous grass moths or crambids in another branch of the same family look large in flight because of their broad hindwings, but seem to vanish when they settle on a stem and instantly furl these beneath their narrow forewings.

Quite a few of the Geometridae are associated with wetlands, but in every case the larval food plant is a herb or a shrub. In Britain at least, no member of this large and varied family feeds on grasses. Even the species whose name indicates that they are associated with grassland habitats have other food plants: gorse and Broom for the Grass Emerald *Pseudoterpna pruinata*, Yellow Rattle for the Grass Rivulet *Perizoma albulata*, and heather for the Grass Wave *Perconia strigillaria*. Ironically, grass was long suggested as the probable food plant of two geometrids when the actual plant was unknown: the Black-veined Moth *Siona lineata* caterpillar was thought to feed on Tor-grass but is now known to eat Marjoram. The natural food plant of the Slender-striped Rufous *Coenocalpe lapidata* is still tentatively given as grasses in many books (e.g. Porter, 1997), but a knowledge of the family shows this to be most improbable. Buttercup, as eaten in captivity, is much more likely.

Many of the geometrids of wetlands, therefore, as caterpillars feed on such plants as bedstraws in the case of the Oblique Carpet *Orthonama vittata*, meadow-rue for the Marsh Carpet *Perizoma sagittata*, chickweed for the Marsh Pug *Eupithecia pygmaeata*, and so on. Where there is carr, especially sallow and Alder, a further range of geometrids will be present, most of which can easily be disturbed from the foliage in the daytime, or found resting on the trunks. The noctuids, prominents and other purely nocturnal species like the Eyed Hawk-moth associated with carr are not often seen in the daytime except in the caterpillar stage. However, it is well worth checking the boles of sallows on sunny mornings for newly emerged clearwings such as the Lunar Hornet Moth (Plate 26).

Plate 73: Drinker Moth *Euthrix potatoria*. Most familiar in its caterpillar stage, the adult can occasionally be found at rest. The female resembles a bunch of yellowing leaves, the palps perhaps being the stalks. (Sussex).

Towns and Cities

It is relatively easy to find moths during the daytime in urban areas. Plenty of moths breed in towns, especially in those parts where there are large mature gardens, tree-lined avenues, and parks. Also, more rural species fly in from the surrounding countryside, attracted by the lights at night. When dawn comes, these moths are at a disadvantage in finding a place to hide. They may be beautifully cryptic against mossy, lichenous bark or rock, but not against man-made artefacts such as a metal light bracket, or the mahogany window frame of a shop. Perhaps, in a few thousand years, some moths will have evolved patterns that conceal them against brickwork, concrete and treated timber fencing – assuming we are not using different materials by then. For the present, the moth hunter has the advantage. Similarly, those species that rest at ground level by day have fewer places to hide. There may be litter in towns, but generally it is not leaf litter, nor does much grass grow on pavements. Often, the best that moths like darts and wainscots can do is to try and secrete themselves in the angle between the pavement and a wall or fence. Fortunately, there are fewer specialist predators to find them in towns: voles, shrews and warblers are much scarcer here than in the surrounding countryside. House Sparrows do eat moths if they come across them, but their attempts to catch them are often clumsy.

Even so, few enthusiasts would deliberately set out to search for moths in towns. It is more of a secondary activity, something to do while on the way to school or work, when shopping, or walking the dog. High streets are often the most productive places, as many premises are brightly lit all night. Gas and electricity showrooms, garage forecourts, estate agents' windows, public houses and late-opening supermarkets attract moths as well as customers, and some will still be there next morning. The best places to look are soon learnt, and it pays to check them as early as possible in the

Plate 74: The Lime Hawk-moth *Mimas tiliae* is quite an urban species, benefiting from avenues of lime or elm in the southern half of Britain. Shape and colour, with a strongly disruptive pattern, conceal it amongst dappled foliage, especially side shoots from the trunks of its food plants. (Sussex).

morning, before the moths are disturbed or eaten by birds. There is no point in being self-conscious or surreptitious while engaged in this activity, as it only arouses greater suspicion. If passers-by ask what you are doing, explain, and show them a moth if possible. Many will be mildly interested.

Certain types of street lighting attract moths. The whiter the light the better, the orange sodium lights are useless. Lamps with mercury-based filaments are excellent. Moths that have settled on the lamp itself will be well out of reach to all but the most determined enthusiast, but any pavements, fences and walls around the light are worth examining. Some large road signs are illuminated at night with powerful floodlights. They also lure many moths, especially if there are no other bright lights nearby, and not all will fly away at dawn. Success is more likely if the previous night was warm, still and damp – suitable conditions for moths to be on the wing.

Also productive are pedestrian underpasses. In effect, they act as moth traps. One I regularly used as a student on the outskirts of Brighton was equipped with mercury vapour lighting, and for several years it provided me with a host of exciting, long-coveted species. Being from northern England, most of them were new to me, familiar only as illustrations in the guides. Alas, the underpass lights were eventually vandalised, and their dimmer replacements behind bullet-proof glass proved nothing like as effective.

Obviously, the range of moths found in a particular town will largely reflect the type and quality of habitats in the surrounding countryside. For me, Lewes in Sussex provides the fondest memories. The diversity of species there was vast. Its lights attracted moths both from the surrounding chalk downland and from the brooks and levels of the River Ouse; it was also close enough to the coast for a sprinkling of scarce migrants. I soon learnt never to go anywhere without a suitable container, and came back from the laundrette with a Cream-spot Tiger, from the fish and chip shop with my first Lime Hawk-moth *Mimas tiliae* (Plate 74), and from my bank with a Lappet Moth *Gastropacha quercifolia*. Yes, towns can be excellent places to find moths.

Chapter 6

Finding moths at night

This chapter is about the various ways to find moths at night as they go about their normal activities, as opposed to luring them to you by means of light or sugar. Why bother, many will ask, when modern mercury vapour light traps are so effective? There are several reasons. Traps are expensive. They need a power source – for the most productive designs either mains electricity or the use of a heavy, noisy, smelly generator (another large financial outlay). Either way, the sites where it is practical to use such a trap are limited. Smaller versions run from heavy-duty batteries, and are more portable, but the catch is greatly reduced. Also, light trapping is selective. Catches are biased. Some moths are strongly attracted to light, others almost ignore it. Traps are less effective for crepuscular species that are on the wing before it is fully dark. As a rule, males are much commoner than females at light, which can be frustrating if the aim is to obtain eggs for rearing.

Worst of all, attracting a moth to a light trap proves only that the species is present in the general area. It tells us nothing about its precise habitat requirements, and rather little about its behaviour. The moth is taken out of its natural environment. For me, there is less pleasure in seeing a moth like that. Imagine if a certain wavelength of light was discovered to attract butterflies. A trap with the appropriate bulb could be taken to a woodland glade at dawn on a May morning, left all day and brought home at dusk. Inside, sitting around on the egg cartons, would be dozens of the commoner species of butterfly and smaller numbers of the scarcer ones. Who would wish to use such a trap? With butterflies, it is the circumstances of the sighting that gives true enjoyment. An Orange-tip *Anthocharis cardamines* resting on a Cuckoo-flower, a Speckled Wood in dappled sunlight on the forest floor, a male Small Copper fearlessly intercepting any other insect that dares to come into his territory, new-minted Red Admirals on deep purple Buddleia

– all these are magical, life-enhancing images. Take away the context, and the butterfly might just as well be a pinned specimen in a collection.

Similar pleasure can be gained from watching moths in their natural surroundings after dark. It takes a bit more effort to observe them in this way, because they are nocturnal and humans are not (we merely prolong the day with artificial lighting). However, it is worth it, for the quality if not quantity of sightings. No other equipment is needed than a torch, and perhaps a net. There is nothing to pack away at the end of a session, and if one spot is unproductive it is easy to move to another. Several different habitats can be covered in the course of one evening. The moths themselves need not be detained, except for any required for photography, breeding or later examination. The rest are allowed to carry on with their vital activities of feeding, mating and egg laying without the disruption of being trapped. And there is usually much of interest besides moths. My favourite site in Sussex had up to 16 singing Nightingales in May, and once I found myself in direct competition for geometrids with a Nightjar (the Nightjar won). The shorter turf was dotted with the eerie green beacons of Glow-worms, easily visible at 30 paces. No one else went there after dark, and the peace was absolute.

Safety

It is sensible to get to know an area fairly well from visiting it in the daytime before going there at night. Make a mental note of any potential hazards and awkward places, as well as areas that look promising. Ensure that you will be able to find your way around – everything looks different in the dark, and it is easy to get lost. Ask permission on private land; this may be wise even if there is public access during the day. Behave openly – moth enthusiasts are regularly mistaken for poachers, who also go round at night shining torches. It is important not to arouse suspicion. Do not lurk or skulk in a furtive manner. Brandish your net like a staff of office, so that everyone can see what you are doing. Lepidopterists are commonly perceived as slightly eccentric, but totally harmless, so turn this reputation to your advantage. Preferably take a companion, or a large, well-trained dog if it makes you feel more confident. Having said this, I have never felt in the slightest danger while mothing in the countryside at night – only when returning on foot in the early hours through the city centre or suburban housing estates!

Dusking

Having chosen your site (which might simply be your own garden), you need to wait for suitable weather. Unless the site is sheltered, strong winds are bad, especially if from the north. Warm, still, humid and moonless conditions are best. On such evenings, walking round with a net at dusk – known to Victorian collectors as dusking – is an excellent way to find moths. A few of the nocturnal species do not fly until the small hours, but most are active from dusk onwards, even though they are not attracted to light until later. Many other moths have strong evening flights, then stop as if choreographed. This may be to do with temperature, which can plummet after dark if there are no clouds to act as an eiderdown. Perhaps also there is a safe slot when most birds (except Robins) have gone to roost, but before the bats come out.

Whatever the reason, there is a period of up to an hour when it is still light enough to walk around easily, but dark enough for moths to be flying. Twilight comes sooner under the shade of the canopy, so it is best to start in woodland, and move to more open areas as the light fades. Geometrids usually make up most of the catch, being easily seen and netted because of their slow, fluttering flight, whereas noctuids travel much more quickly and require fast reactions. Dexterity and timing improve with practice. I prefer to use a white rather than a black net when dusking, finding it much easier to see the captured moths – and easier to find the net if it is put down.

At first, it will be light enough to identify the catch without using a torch. Some of the carpets, however, have the frustrating habit of not opening their wings, thus hiding the diagnostic upperside pattern. The Red Twin-spot Carpet *Xanthorhoe spadicearia* is one that does this so consistently that it almost becomes an identification clue in itself. With experience, many of the commoner species can be recognised even by their undersides, or from their appearance in flight. Netting then offers the opportunity to be selective, picking out moths that look slightly different or unusual rather than trying to catch every one that is encountered. As the light fails, a torch will be necessary to help identify the catch. Unfortunately, this has an adverse effect on night vision, making it harder to see moths to net. Once the torch is required for this purpose too, it is usually time to stop anyway, and either go home or turn to other techniques. Dusk is the best time for dusking.

Natural Attractions

Flight is energetic, requiring powerful muscles to overcome the pull of gravity, and in turn these muscles need fuel. Once airborne, long-winged birds like shearwaters, gulls and raptors can glide and soar, saving energy. Some butterflies can glide too, like the White Admiral *Limenitis camilla* and other members of the Nymphalidae, but no moths appear to do so, and most are the wrong shape anyway. On emergence from the pupa, all moths are provisioned with stores of fat – lipids – laid down by the caterpillar, to provide energy. For some moths, including whole families like the eggars, prominents and tussocks, these stores are sufficient for the rest of its life. Lacking a functioning proboscis and digestive system, the moth cannot feed, and shows no interest in flowers. When its fuel runs out it dies. Such species cause difficulties for the collector, as any unused lipids gradually permeate through the entire specimen, making it greasy.

A commoner strategy is for moths to be able to augment their reserves by feeding on nectar, which is an excellent provider of energy, or other plant juices. The water content may be just as important, to prevent dehydration. Nearly all members of the two largest families of macro-moths, Geometridae and Noctuidae, are able to feed, although a few have lost the ability, like the Peppered Moth and the Sprawler *Brachionycha sphinx*. The latter flies in early winter, when there are few nectar sources anyway. Curiously, a full-sized proboscis case is present in its pupa, but remains empty during development, showing that the loss is secondary and possibly recent in evolutionary terms. Male Winter Moths, which fly at the same time of year, have a much reduced proboscis but still use it to take in moisture.

Given that so many adult moths do feed, this behaviour is a useful aid to finding them. Like butterflies, moths visit flowers, but they also utilise a rather wider range of food sources. For the moth hunter, it is both fortunate and convenient that there tends to be one predominant attraction at any particular time of year. As that attraction fades, and loses its allure, another takes its place. By working each in turn, it is possible to find good numbers of feeding moths from March right through to November, at least in southern England.

It is often claimed that modern urban man, living and travelling in an increasingly sterile, air-conditioned, artificially lit environment, has become divorced from the natural rhythms and seasons. An interest in observing moths is the perfect antidote. Like some Medieval peasant, you will reckon your year not solely by the date on the calendar, but by whether the sallows are in bloom yet, or the blackberries ripe. It will actually matter to you what time it gets dark. Like the peasant, you will become attuned to every nuance of the weather – a minor change in wind direction, the percentage of cloud cover, a rise in temperature – and the phases of the moon. All might portend a successful evening, or a pointless waste of effort, and it is preferable to know which.

Each habitat, and each site, will have its own mix of seasonal attractions. It might take several years of fieldwork to learn what they are, and when best to work them, but this accumulated local knowledge is carried forward into the future. Certain major attractions, however, are so widespread and so productive that they deserve to be considered in further detail.

Sallow Blossom

This is the first great feast of the year for moths. In southern England, sallows are often in bloom by the beginning of March, but not until mid-April in northern Scotland. By day, the 'pussy-willow' catkins hum with bees, but seem relatively unattractive to butterflies except for the occasional Small Tortoiseshell. At night it is a different matter. With few rival nectar sources at this time of year, they concentrate most of the moths that are on the wing in early spring. These fall into two groups. First, there are those species that pass the winter as adults. Many will have been fasting for several months, and urgently need to refuel. Some hibernating species are unable to mate or produce eggs until they have fed properly in spring. Secondly, there are moths that emerge in March and April, having overwintered as pupae. Most numerous are the Quakers in the genus *Orthosia* (Plate 75), several of which are abundant throughout most of Britain. Even some early geometrids like the Early Thorn *Selenia dentaria* (Plate 15) and the Engrailed, not often seen feeding, occasionally attend.

Sallows are dioecious (producing either male or female catkins on separate plants). Moths find both sorts equally attractive, which is hardly surprising if the moths are to act as pollinators. They arrive at dusk, quickly settling on the bushes, and feed quietly for an hour or two, crawling from catkin to catkin rather than flying. Because they do not need to fly much, they are less daunted by adverse weather conditions. Moths can be found at sallow even on nights that are too windy, rainy or cold for dusking, or for operating a light trap with much hope of success.

Plate 75: Large numbers of moths nectar at sallow catkins after dusk, like this Twin-spotted Quaker *Orthosia munda*. (Sussex).

Where the sallows are low enough, it is a simple matter to walk round the bushes with a torch. This should be reasonably powerful, but not so bright as to dazzle the moths and cause them to release their hold. Moths are perhaps easier to see against the fluffy, golden male catkins rather than the spiky, grey-green female ones. Sometimes their eyes reflect the beam like tiny jewels. However, often the sallows (especially Goat Willows) are too tall, and the moths are way out of reach overhead. Here, the recognised technique is to give the boughs or trunk a sudden jar, causing at least most of the noctuids to drop to the ground, together with a rain of spent catkins. In bygone times one's servant used to spread a sheet on the ground first, so that the moths would show up better. A similar result can be achieved by brushing away any leaf litter and debris from under the tree.

Much of the enjoyment of working the sallows comes from seeing moths in such numbers again after a winter's dearth. Nor is it necessary to stay out very late, as dusk comes early at this time of year. Admittedly, most of the expected dozen or so species are common, but it is nice to see old friends again, and there is always a chance that one of the scarcer hibernators will turn up. The scent of sallow blossom hanging in the calm damp evening air still brings back the excitement of my first Tawny Pinion.

Campions

Many of the campions and catchflies in the genera *Silene* and *Lychnis* depend primarily on moths to act as pollinators. Some have evolved specialised features to try and ensure that their nectar is available only to the chosen few: the flowers of the rather local Nottingham Catchfly look dead and shrivelled during the day, but open at dusk, as do those of the Night-flowering Catchfly. Presumably their scent is also formulated to appeal especially to moths.

Certainly, campions in general are a favourite source of nectar. Noctuids in the genus *Hadena*, with nine species on the British list (including the recently extinct Viper's Bugloss) specialise on them. They exact a high price for acting as pollinators, as their caterpillars feed mainly on the developing seeds, sometimes eating the contents of every capsule on the plant. To minimise inter-specific competition, each moth prefers one species of campion or habitat, as Young (1997) points out, although there is some overlap. The Campion *Hadena rivularis* is associated with Ragged-Robin, the Lychnis *H. bicruris* with Red Campion, and the Marbled Coronet *H. confusa* with Sea Campion. Incidentally, each of these species has a partial second brood in the south of England, unusual for any moth whose caterpillar feeds on seeds. However, the long flowering season of campions, which often 'come again' in August, makes this possible, although they are at their best in May and June.

Besides the *Hadena* group, many other moths regularly nectar at campions. Whereas at sallow the moths sit on the catkins, on campions most employ a different technique. They hover from flower to flower, resting briefly with quivering wings to feed, then moving on. Therefore it is often necessary to net them for identification. However, darts in the subfamily Noctuinae are an exception. They always settle with closed wings on the campions, and do not hover.

At any site, the precise species mix depends on habitat, but most will be noctuids. Red Campion grows on fertile, non-acid soils, often in woodland glades and rides or along hedgerows. In such places there are likely to be nettles too, the food plant of plusias such as the Burnished Brass *Diachrysia chrysitis*, Beautiful Golden Y and the Spectacle *Abrostola tripartita*, all avid nectar drinkers. White Campion is an annual, associated with disturbed ground like the margins of arable fields where the sprayer has missed and the sides of farm tracks. Here it will be used as a nectar source by moths whose caterpillars feed on other weeds of crops; for example the Shark *Cucullia umbratica* which feeds on sow-thistles, the Broad-barred White *Aetheria bicolorata* on hawkweeds.

Bladder Campion is more local. It prefers light soils with a high pH, so is common on chalk downland with disturbed ground such as quarries, road embankments and field headlands. It also thrives on base-rich industrial slag heaps. In Sussex, large clumps grew on one chalk bank in the excavated earth of old Badger setts. By day, there were Adonis Blues and Dingy Skippers *Erinnis tages*, but after these butterflies had gone to roost it was equally pleasant to sit there listening to the Quail and watching the Small Elephant Hawk-moths, so close that the hum of their wings could clearly be heard.

Most of the activity at campion takes place in the hour after dusk. As soon as they get on the wing, moths replenish their food reserves, and probably their water requirements as well after a hot day. Having done so, they turn to other activities like dispersing, mating and egg laying. Once it is fully dark, few will be seen feeding, so working campion is best treated either as a short evening session, or as a prelude to light-trapping. It can easily be combined with dusking, thus sampling the geometrids and micros too.

Rosebay Willowherb

The coming of the railways helped to spread this once-local perennial throughout Britain. It is quick to colonise newly burnt ground. Sparks given off by coal-fired locomotives regularly set fire to railway embankments and cuttings, providing this plant with an ideal opportunity, while the slipstreams of passing trains carried its silky pappused seeds in their wake. Multiplying also by means of underground rhizomes, now it often forms dense stands along road verges and beside woodland rides.

Like many other plants that attract moths, Rosebay Willowherb is not particularly favoured by butterflies, except for the whites *Pieris*. In the south of England it is at its best in July, a time when there are few other large-scale natural attractions. In Scotland, August is its peak month. The moths attracted reflect the quality of the surrounding habitat. In ordinary farmland abundant and generally distributed species predominate, like the Smoky Wainscot and Common Rustic. However, where the plant grows in woodland a more varied and interesting selection can be found. For instance, it is a favourite of the Triple-spotted Clay and Slender Brindle *Apamea scolopacina*, moths that are hard to find by other methods. As well as noctuids, geometrids too are able to reach the nectar of Rosebay Willowherb. The Small-fan-footed Wave *Idaea biselata*, and the second brood of the Small Phoenix *Ecliptopera silaceata*, for which it is the larval food plant, can often be seen in large numbers. The July Highflyer often abounds, as the old authors would say, likewise the Dark Marbled Carpet in Scotland. They always feed with wings closed, butterfly fashion, making it frustratingly difficult to check for interesting forms of such immensely variable species. Most of the Larentiinae behave like this, but one exception is the Pretty Chalk Carpet *Melanthia procellata*, which feeds with wings open.

Rosebay Willowherb commonly occurs in linear stands by the side of tracks. In Sussex, it grew along a bridleway through a scrub-filled downland coomb, for a distance of about a mile. A leisurely stroll with a torch, checking the plants on one side on the way out and the other on the way back was a pleasant way of seeing some interesting moths. There was no need to repeat the walk: as with all natural attractions, the hour after dusk was best. The willowherb was tall and vigorous, so the moths were more or less at eye level, and showed up to best advantage. Although there was no one there to hear it, I could not avoid an involuntary exclamation of delight on seeing the bright apple-green of a female Scarce Silver-lines (Plate 12) against the shocking pink of the flowers. The same moth in a light trap would never have provided so vivid and memorable a view.

Honeydew

Aphids suck in the sap of plants, and continuously excrete the surplus sugars and water from special glands. Ants often guard aphids, and milk them for this sweet fluid; what is left forms a sticky deposit known as honeydew on the upper surface of leaves, and this is very attractive to moths. As the gardener knows, there are many different aphids, each usually confined to one species of plant.

Unfortunately for the lepidopterist, the quantity of honeydew is unpredictable. It depends partly on the fluctuations of the aphid populations, and partly on the weather. Rain washes off honeydew. A settled warm dry spell is therefore best, both to encourage the aphids to multiply and to ensure a build-up of their secretions on the leaves. Drought also serves to make moths thirsty, yet at the same time reduces alternative sources of food and moisture such as nectar. Too sticky in sunshine for butterflies, the honeydew becomes available to moths once dissolved in the film of dew that so often condenses on leaves when a calm clear evening follows a scorcher of a day.

In some years, honeydew will not figure at all in the lepidopterist's activities. Unlike the other natural attractions, its occurrence is not governed by the calendar, except that June to September is the most likely period. Nor is past experience much of a guide: honeydew tends to be a 'one-off' event. The best I ever encountered resulted from aphids on Stinging Nettles growing along a fence line at a downland site in Sussex. While scrambling through the fence one evening in early June, I noticed a couple of moths on the nettle leaves, and soon found many more. For the next 3 weeks the weather stayed hot and dry. The honeydew built up, and so did its clientele of moths after dark. The same distinctively patterned or scarred individuals could be recognised night after night, often feeding within a yard or two of where they had previously been seen, gradually becoming more and more worn. On the best nights there were literally hundreds of moths. The variety was astonishing. Unlike nectar, honeydew is accessible to all species whatever the length of their proboscis. The biggest and smallest moths present, the Large Yellow Underwing and Haworth's Pug, sometimes fed on the same leaf, an amusing contrast. In all, I recorded 60 different macro-moths, 31 of which were noctuids and the rest geometrids apart from the Peach Blossom. They included several that are rarely witnessed feeding: the Rosy Marbled, Straw Dot *Rivula sericealis*, Small Waved Umber and Clouded Border. Three weeks later the show was over. Cattle ate the sugar-coated nettles, the weather broke, and the aphid population crashed. The phenomenon was never repeated. In subsequent years, as in previous ones, there were few aphids on the nettles and even fewer moths.

Sycamore, a much maligned tree, is slightly more reliable as a host of aphids and source of honeydew. Its large flat leaves provide good platforms for feeding moths, and also make them easy to spot. While most of the honeydew and therefore moths is out of reach, some does drip onto the lower branches and any vegetation growing beneath. My first

Plate 76: Like other autumn noctuids, the Orange Sallow *Xanthia citrago* visits honeydew, especially on the leaves of its food plant, lime. At rest amongst fallen leaves, disruptive lines break up its shape. (Banffshire).

encounter with honeydew was during a boyhood Indian summer (they seem to have been commoner then!). Torchlight revealed numerous moths on Bracken fronds in a wood, apparently feeding. Bracken itself provides no nourishment for moths, and it took me a while to work out that honeydew from the aphid-infested Sycamores overhead was the attraction. As it was early October, many of the autumn moths were present, including the Brick and the Yellow-line Quaker which were new to me at the time, however commonplace now.

There is little information about moths and honeydew in the literature. Even Tutt (1901–05), giving practical hints for fieldwork, mentions the subject only once, and briefly. No geometrids are given as attending honeydew in either South (1907–09) or Skinner (1998), yet some of them I have never seen feeding anywhere else, for example the Scalloped Hazel. However, moths such as the *Hadena* group and the Plusiinae that normally feed by hovering in front of flowers are not attracted to honeydew, just as they rarely visit 'sugar'.

Ragwort

Common Ragwort is a rather undesirable plant. Full of alkaloids, it is highly toxic to grazing animals, especially horses, causing progressive and fatal liver damage. For this reason, it is one of a small group of noxious weeds that landowners have a statutory duty to destroy lest the seeds spread to a neighbour's land, though this law is hardly ever enforced. Nor is its scent attractive, causing it to be nicknamed Stinking Willie in some parts of the country.

None the less, it is a good nectar plant for moths. A few, like the Antler Moth (Plate 77) and *Udea lutealis* can often be seen there in the daytime, but there are many more in the hour or two after dark. Flowering mainly in August and through September, ragwort conveniently follows on from Rosebay Willowherb as the chief natural attraction in many areas. Its several species are quick to colonise disturbed soil, being characteristic plants of overgrazed pastures, roadsides, waste ground and derelict industrial sites. Wherever it grows, ragwort will attract moths, but obviously it can only attract the ones that are there. This means that in degraded habitats such as intensive farmland, the

Plate 77: Ragwort heads are a major source of nectar for moths. Some, like this Antler Moth *Cerapteryx graminis,* can even be found there during the day. (Banffshire).

most ubiquitous and commonplace species predominate. In such areas, Common Rustic, Lesser Common Rustic and Square-spot Rustic *Xestia xanthographa* will often make up over 90 per cent of the moths seen, but at least they are very variable, having a wide variety of different forms. And ragwort is easy to work. Moths show up well against the flat, yellow flowerheads, conveniently between knee and waist height. So if the results of a session should prove disappointing, at least there is the consolation that not much effort has been expended.

Where ragwort grows on chalk downland, there is a chance of more unusual finds, except that so many of the specialities of this habitat are diurnal, or on the wing before the plant is at its peak. Woodland glades offer better possibilities. In Sussex, I used to find the Tissue on ragwort, looking as big as a butterfly as it fed up prior to hibernation. There were also pyralids that even I could identify, like *Evergestis pallidata*.

However, ragwort really comes into its own at the coast, especially as normally there are few other major sources of nectar there. It grows well on sand dunes, and on disturbed ground below sloping cliffs, often flowering slightly earlier here than inland. It offers an excellent way to see those littoral darts that are nigh impossible to find by day, as well as large numbers of other mainly coastal moths such as the Rosy Minor *Mesoligia literosa*. Geometrids like the Annulet are attracted too, and of course there is always the chance of scarce migrants, especially in September.

Blackberries

Blackberries are the fruit of the numerous microspecies of bramble, of which there are about 400 in Britain. Some have more luscious and palatable berries (drupes, to the botanist) than others, but overall they provide an abundant source of autumn fruit exploited by many creatures. Humans, deer, foxes and mice relish them. Given the choice, birds prefer elderberries, which are more easily swallowed (Snow and Snow, 1988), but then turn to blackberries. Among the insects, numerous flies and wasps ingest the juices of the over-ripe berries, but these attract few butterflies apart from the occasional Red Admiral and Comma. However, numerous moths can be found on blackberries after dark, especially in woodland.

As with other natural attractions, a prior daytime reconnaissance is worthwhile, to ensure that the quantity is sufficient and the timing is right. The sheltered edge of a wood, or an open ride or glade, offer the greatest possibilities. It is best to wait until much of the fruit is over-ripe. Many of the more interesting moths that can be found at blackberries do not emerge until late September anyway. Superstitious country people once believed that the devil flew abroad after Michaelmas, spitting on brambles to turn the fruit mouldy (Hudson, 1923). Moths actually prefer blackberries in this state, slightly fermenting and decomposed. The Herald moth (Plate 44) has a barbed proboscis able to pierce the skin of soft fruit, but other species can reach the juice only if the fruit is damaged.

Most of the moths found at blackberries will be noctuids, especially those presently included in the subfamily Cuculliinae. This includes many species that fly in the autumn and overwinter in the egg stage, like the Green-brindled Crescent, Merveille du Jour, and the various members of the genera *Agrochola* and *Xanthia*. Others in the subfamily also emerge in autumn, but hibernate as adults. Before doing so, they spend much time building up their resources while expending the minimum of energy, so blackberries are particularly important to them. The two commonest members of this group are the Chestnut and the Satellite; both will be seen again at sallow blossom in the spring. Others in this category are scarcer or more local, and as they do not fly much in autumn they tend not to be caught in light traps. Blackberries

Plate 78: Ripe blackberries are an important food for autumn moths, especially those that hibernate as adults like this Red Swordgrass *Xylena vetusta*, now mainly northern and western in Britain. (Banffshire).

may offer the best opportunity of seeing fresh specimens of the Pale Pinion *Lithophane hepatica* and, where they still occur, the Sword-grass and the Red Sword-grass *Xylena vetusta* (Plate 78). Geometrids are few, but females of both the Autumn Green Carpet and the Red-green Carpet regularly attend blackberries at my Banffshire site.

Moths at blackberries are usually docile. They arrive at dusk and drink their fill for an hour or two from the same spray of fruit. If the night cools rapidly, a film of condensation may form on their thorax and forewings, like breath on a windowpane, yet they carry on feeding. A net is not required, and may indeed prove a hindrance as it snags on the thorns. To catch a moth for closer examination, hold a jar or pill-box directly below it, and cause it to fall in by gently tapping the plant.

Autumn moths are often dressed in autumn colours, to blend with the fallen leaves. Their rich warm shades of orange and yellow, russet and chestnut seem to glow in torchlight against the ink-black berries. All around, the mouldering, fermenting, fungoid scents of autumn hang heavy in the moist night air. Perhaps the first migrating Redwings will be heard overhead. Make the most of this time of year, for soon it will be winter.

Ivy bloom

The globular, yellowish-green inflorescence of Ivy hardly merits the term blossom, but it does produce nectar at a time of year when it is in short supply. Only plants growing in sunlight, as along the very edge of a wood, will flower, but often they do so copiously. The blooms appear in October and continue into November, making Ivy the last great natural attraction of the year for moths.

At first, Ivy bloom competes with ripe blackberries for the moth hunter's attentions, and indeed most of the same species can be found at each. Possibly more geometrids are seen at Ivy, like the November Moth and Pale November Moth. Some time in October, the blackberries are over, and only Ivy is left. Little is on the wing apart from the hibernating species and the Angle Shades, a moth that can be seen in every month of the year. However, it is still worth checking Ivy bloom, especially near southern coasts, in case unusual migrants turn up. One evening in late October, the first moth I found was a Scarce Bordered Straw *Helicoverpa armigera*. So was the second – but that was the end of the excitement.

Plate 79: Often the last species of the year, mating pairs of Winter Moths *Operophtera brumata* are easily picked out in torchlight after nightfall, providing a remarkable example of sexual dimorphism. Like many geometrids, the male holds its wings in the butterfly position at night, and the newly emerged female likewise has her own stubs raised as if drying them. (Banffshire).

Miscellaneous

At night, torchlight will reveal many moths. They can be seen sitting about on vegetation between flights, drying their wings if they are species that emerge from the pupa after dark, or found as mating pairs. Most moths show up clearly, and can often be spotted at much greater distances than they could in daylight. This is because the scales on their wings, especially the undersides, reflect the directional light of a torch beam like road signs caught in car headlights.

Wandering about with a torch is a particularly effective technique from late autumn through to early spring. Several geometrids are out at these seasons, and the males sit openly on the leafless trees and bushes or on withered vegetation beneath, especially if the night is calm. None of the species concerned is particularly rare, and several of them are abundant, like the Winter Moth (Plate 79) and the Mottled Umber *Erannis defoliaria* in all its attractive diversity of colour schemes and patterns. One of these is almost certain to be the last species on the annual list, yet there is something special about moths that can fly in the dead of winter when the ground crackles underfoot with frost. I never tire of ending the year with at least one excursion to see them on the leafless twigs at night. Similarly, the first Pale Brindled Beauty or Early Moth *Theria primaria* on the Hawthorn hedges after dark signals the welcome beginning of a new year, often as early as January. Then it is time to look forward to the sallow blossom once again.

Chapter 7

Lures and traps

The use of lures and traps gives the lepidopterist some control. They bring moths to the observer, whereas with other types of fieldwork the observer goes to the moths. It is still important to choose a suitable habitat in which to operate: obviously, the desired species must be present in the general locality, and within range of the attractant. However, sometimes the ability to lure moths even a few metres can make all the difference, for instance when the precise area where the moths are flying is too dangerous or fragile to work directly, especially at night. Fens, marshes and reed beds, old quarries and sea cliffs fall into this category. Unfortunately, there are also many tales of unscrupulous collectors setting up their pitches just outside the boundaries of nature reserves where they are forbidden to operate, and enticing rare species to their doom.

Sugar

It is not surprising that moths are attracted to sugar, since this is only the refined and crystallised sap of plants, which they already consume naturally in the form of honeydew or directly from wounded trees. But whereas honeydew is sporadic and unpredictable, and nectar sources too can be seasonal and unreliable, sugar can be used anywhere and at any time of year. Whether or not it proves successful is a different matter.

Recipe

The concoction known to lepidopterists as sugar is easy enough to make. A basic recipe requires the following ingredients, available at any supermarket:

1. One tin (454 g) of black treacle
2. 1 kg of muscovado sugar
3. 500 ml of brown ale

Pour the brown ale into a very large saucepan, and slowly bring to the boil. (It is important to get rid of all the carbon dioxide and alcohol before adding the other ingredients, otherwise the mixture will overflow from the pan like lava from a volcano.) Next, add the muscovado sugar, and stir with a wooden spoon until it is completely dissolved. Finally, mix in the black treacle, bring back to the boil, and simmer for a minute or two. For safety, allow to cool before decanting the result into tins – the ones the treacle comes in are perfect. Never use glass jars or bottles, as these can explode if fermentation takes place in hot weather. Then wash everything up, to preserve domestic harmony!

Much mystique surrounds the making of sugar. Fowler's is claimed to be the best brand of treacle, but seems to be no longer available; Newcastle Brown is said to be the most suitable ale. Additional ingredients, often secret, are legion. Some people add honey, or rotten bananas, or concentrated fruit juice in autumn. A tablespoonful of rum is usually stirred in just before use, and a few drops of amyl acetate. However, I much prefer not to put in anything that might conceivably harm my intended guests or make them tipsy. Tutt (1902) describes moths becoming so intoxicated by the rum that they topple off the sugar patch, to be eaten by frogs and mice waiting below in anticipation of a nightly meal. That seems a little unkind to the moths. Then there may be other reasons for keeping the sugar wholesome. On several occasions, far from home and staying out later than planned on an exceptionally good night, I have been more than glad to consume what was left in the tin myself!

Using sugar

Sugar is most commonly applied to tree trunks or fence posts, just before dusk. Where these are absent, it can be smeared on any convenient tall vegetation: bramble leaves, the flat umbels of Hogweed, reed stems tied together in a bundle, or whatever else is available. It is best to apply a good dab of sugar at about eye level on a tree trunk, or near the top of a fence post, because it will slowly trickle down of its own accord. A 1½″ paintbrush fits nicely into the treacle tin. About 25 trunks or posts, spread over a distance of about 100 metres, makes a reasonable beat for one evening.

Where and when it is best to use sugar are difficult questions to answer. Allan (1947) entertainingly goes into considerable detail on the sort of locations and weather conditions to avoid, compared with those that offer the best prospect of success, but there are no hard and fast rules. It is easy to become disillusioned after a few disappointing sessions, and jump to the conclusion that sugar is a waste of time in one's area. Part of the problem is that the reason for any lack of moths is not always clear. The site itself could be ill-chosen, or the weather unfavourable. Even if both are perfectly suitable, competing natural attractions such as honeydew might be too strong on that particular night. Many sessions, spread over different times of the year, can be necessary before the final verdict is reached.

However, it is safe to give a few general guidelines. Moths dislike wind, but need uncluttered airspace in which to fly. Therefore, choose a sheltered yet open area. The middle of a wood is universally held to be poor for sugaring, unless there is a broad ride or large glade. Otherwise, the lee edge of the wood is to be preferred. Some hollows, dips and valley bottoms can be *too* sheltered on still nights, when cold dank air sinks and accumulates; a little higher up the slope the temperature may be several degrees warmer, so avoid areas where the morning dew is often heavy. The boundary between two different habitats is always worth trying, if only because the sugar might then attract moths characteristic of both, thus widening the range of potential species.

Following these hints is no guarantee of success. In fact, my own favourite sugaring site – a low bank separating wet heath and carr from mixed farmland – was found purely by chance. Setting out to work what seemed a promising spot a mile away, I wiped the brush on a few nearby fence posts to remove sticky traces from the previous session before packing it in the rucksack. Returning a couple of hours later, having seen hardly anything at all, I shone the torch on the posts used for cleaning the brush. They were covered with moths! Best of all, this fence was only just across the road from the house. Convenience is a great asset: a moderately good sugaring site worked every suitable evening will produce more moths (and more useful data) than a better one that is too far away to be worked regularly. As with honeydew, the clientele builds up night after night, and there is a saving on materials too. After nearly 10 years, my own posts are well impregnated and need only be topped up with a small dab of fresh sugar before each session.

As always, weather conditions play a part. Warm, humid, calm nights are best, though a gentle breeze to waft the scent of the sugar out across the neighbouring habitat does help. Winds from the quarter between north and east usually presage poor results, probably because such nights tend to be too cold anyway. Balmy south-easterlies, on the other hand, quicken the anticipation, as they may bring scarce migrants. Yet some apparently ideal nights produce very few moths even at a tried and trusted location. Usually there is no obvious explanation, though often the weather deteriorates soon afterwards. It has been suggested that moths can detect changes in barometric pressure, forewarning them of this. Conversely, moths can still be found at sugar when it is too showery, windy or cool to make light trapping worthwhile.

As with the natural attractions, moths come most strongly to sugar in the early part of the night, especially the hour after dusk. Having drunk their fill, they leave. Even on a good night, there is little point in making more than three rounds, about half an hour apart, to check the sugar. The second round is normally the most productive, both for numbers of individuals and variety of species, whereas the third round simply reveals the same, but fewer, moths. This rule does not apply when immigration is actually in progress; then, moths continue to arrive throughout the night. They are often hungry and feed greedily. On such special occasions, it is obviously worth continuing the rounds for as long as they are profitable.

Nearly all the moths that feed at honeydew will come to sugar, and the species mix is very similar. Almost invariably, noctuids make up the great majority of individuals. The Geometridae attend only in ones and twos, but may include such unlikely species as the Barred Red *Hylaea fasciaria*. Even among the noctuids, attendance at sugar does not necessarily reflect their relative abundance in the area. Darts (Noctuinae) are disproportionately numerous except for the True-lover's Knot, which much prefers heather bloom. Similarly, moths in the genera *Cucullia* and *Hadena* are predomi-

Plate 80: On the best nights, sugaring can be remarkably effective. Here moths are literally fighting for dining space – although all are common and ubiquitous noctuids, including four species of *Apamea*. (Banffshire).

nantly nectar feeders, but the Campion will come to sugar if its favourite Ragged-Robin is wet from rain. The Plusiinae likewise are flower specialists, although the Silver Y is a regular attender in autumn when nectar sources are scarce.

Sugaring is indeed most productive when there are few competing natural attractions. At my own site in Banffshire it will attract one or two species even on mild January nights, and is very effective in spring until the middle of April. After that, moths desert it for the sallow blossom. Once the sallows are over there is a resurgence of interest. On cloudless midsummer nights sugar will out-perform a light trap, because at this latitude the nights are very short and for about a month either side of the solstice there is no proper darkness. In good years there can be 300 moths at once on the 25 fenceposts, jostling for room (Plate 80). The Dark Arches is particularly aggressive in these situations, running at other moths and shouldering them aside. Within an hour the sugar is sucked dry. Then at the beginning of August the heather blooms, drawing many moths away. At least it is dark enough now for a light trap to be effective. However, sugar really comes into its own in autumn. There are few blackberries in my area, and most of the flowers are over. Autumn noctuids tend to be sluggish, flying little and so not often coming to light, but they feed greedily at sugar. Micros are numerous too: often half a dozen different species of *Agonopterix*, stoking up before hibernation, as well as many tortricids including *Acleris emargana* with its unique forewing shape. A similar pattern has been observed in Shetland, where sugar is not only more productive than a light trap in September, but also attracts a completely different range of species including scarce migrants (Pennington, 1996).

Sugaring, then, should be seen as complementary to other methods of finding moths. Used consistently, it helps to give a more rounded picture of the species present in an area and their true status. My own sugaring ride approaches within 25 metres of my light trap but the two catches are often very different, even considering only those species that do come to both. If relying solely on data from light trapping I would consider the Peach Blossom (Plate 81) to be scarce here (occasional singles, but not every year), yet at sugar there are sometimes three on one patch, a lovely sight. Several noctuids are consistently 10 times more numerous at sugar than at light: the Double Dart *Graphiphora augur*, the Swordgrass and Red Sword-grass (before hibernation), the Brick, Pink-barred Sallow *Xanthia togata* and Haworth's Minor *Celaena haworthii* among them. Conversely, the Autumnal Rustic and Rosy Rustic are rarely seen at sugar, but are abundant in the light trap. Clearly, over reliance on just one method of recording can give a false impression.

Plate 81: The Peach Blossom *Thyatira batis* is one of the most attractive, early and skittish visitors to the sugar patch, but is less strongly attracted to light. (Banffshire).

There are other advantages of using sugar. For photography or as specimens in a collection, moths ideally should be in pristine condition. At the height of the summer, so many insects enter a light trap on a good night that, instead of settling down quietly on the egg trays, they constantly disturb each other. Unless the desired moths are noticed and boxed immediately, which is not always easy in the harsh glare of the mercury vapour bulb, most inevitably become rubbed to some extent. If left until morning, it is rare to find anything worth adding to a collection or photographing in a catch of several hundred. At sugar, on the other hand, moths hardly ever damage themselves. Surprisingly, they never seem to get stuck, or soiled. On a few nights (perhaps when they can hear bats nearby) all the moths are very nervous and alert, but normally they are quiet enough to examine individually and at leisure. Any required can be nudged into a pillbox or jar, and the others left in peace. With practice, this operation is easily conducted one-handed, while holding a torch in the other. The container should be held directly below the moth, which is then lightly touched with the index finger. Nine times out of ten, the moth obediently drops into the jar. On the tenth occasion the moth flies away or falls into long grass, and inevitably it is always a particularly exciting species. Perhaps the rush of adrenaline affects dexterity, just as binoculars take longer to focus whenever a potential rare bird comes into view. Fortunately, the lost moth often comes back to the sugar patch. Once I was convinced I had fumbled a Lunar Yellow Underwing *Noctua orbona*, which would have been my first, only to find on the next round that it was merely an undersized Large Yellow Underwing.

It is not only moths that come to sugar. Occasionally a precocious caterpillar, keen to sample adult pleasures, will drink the sweet fluid. Caddis flies and crane flies, earwigs and strange beetles, slugs and harvestmen all claim their share. The Larch Ladybird and the Eyed Ladybird are regulars at my site. Wood Mice and Bank Voles climb the fence posts, and probably eat some of the moths too. Robins with their large, liquid eyes, more crepuscular than other birds, take a toll of the early arrivals until darkness falls. Roe Deer tiptoe shyly to lick, then crash noisily away through the undergrowth at the start of a new round. In the daytime, flies swarm, and if there are any Red Admirals about they often visit the sugar too. Once I counted 18, each on its own patch. No other butterflies seem at all interested, though there is always hope of a Camberwell Beauty.

Wine roping

This technique, a variation of sugaring, was introduced by Goater (1986) and popularised by Waring (1995a). A thick cord or light rope, made of natural rather than synthetic fibres so as to be absorbent, is soaked in a vat of cheap red wine and sugar. Cut into lengths of a metre or so, the rope is then draped over low tree branches, bushes or fences. As the ropes can be collected up at the end of the session, to be soaked again and re-used another night, it is more economical than the normal method of sugaring.

Pheromones

If you have bred a female moth (assuming that she has emerged at the right time of year), or been lucky enough to find an unmated female in the wild, she can be used to attract a male. This process is known as assembling. Most moths mate at a set time of day or night. Usually the female begins to release her pheromones as soon as her wings are dry, or at dusk in the case of nocturnal species that emerge in the daytime. The results can be spectacular. They are most easily observed in the large, day-flying species like the Emperor and Kentish Glory (Plate 82), a fact exploited by greedy Victorian collectors. Males can be seen approaching upwind, sometimes losing the scent trail in a capricious breeze, then tacking erratically from side to side until they find it again. It reminds me of the children's game of Hunt the Thimble: warmer, warmer ... colder, colder .... Often the last few centimetres prove the most difficult, especially if the female is partly concealed by vegetation; perhaps her scent is then all-pervasive and the male has to locate her by sight. Once the two moths mate, the female stops 'calling', and no more males arrive.

Assembling can also be employed with nocturnal moths – sometimes unwittingly. Several times I have released a bred female in suitable habitat, and found her still in the same place next morning, but conjoined with a male (Plate 83). This only occurs in species that remain coupled for most or all of the following day, like many prominents and some hawk-moths.

Plate 82: Kentish Glory *Endromis versicolora*, mating pair. The much smaller male is clinging solely to his mate, a rather deeply coloured female. Possibly her hairy abdomen (Plate 9) helps him gain a foothold. This is one of few moths that can be easily sexed as full-grown caterpillars, purely by their size. (Moray).

For a few moths that are serious pests, it has been economically worthwhile to identify and synthesise the specific pheromones released by the female. In Britain, those of the Pine Beauty are used to attract males into standardised traps containing a killing agent. This is not in itself intended as a means of population control, as no females and only about a third of the males present are caught. Instead, the traps accurately monitor population levels, so that forestry plantations need be sprayed only in those few

Plate 83: This bred female Iron Prominent *Notodonta dromedarius* (upper moth), released on Alder at dusk, was found paired with an equally fresh male next morning. Both are of the darker, northern form. (Banffshire).

years when numbers rise damagingly high. Obviously, this is both environmentally and commercially sound.

Not unsurprisingly, for male moths the female's pheromones are the most powerful of all attractants. After 3 years of intensive sugaring, light trapping and daytime fieldwork I was convinced that the Pine Beauty was absent from my home patch. David Barbour, who has worked professionally on this species for many years, was equally confident that it would be present, and lent me three of his pheromone traps (without the killing agent) to hang from the branches of my few stunted pines. Sure enough, several males were caught, but for once I was delighted to have been proved wrong – it was a new species for me.

There is scope for using this technique more widely. It has been tried on a small scale, so far without success, in attempts to discover additional colonies of the New Forest Burnet on the Scottish sea cliffs, and to find new sites for the very local Rannoch Sprawler. Perhaps assembling, using bred females from foreign stock, would be the best way to ascertain whether or not the Small Lappet still survives on English moorlands.

Light trapping

Even those least interested in natural history know that moths are attracted to artificial light at night. Presumably this was discovered when humans first learnt how to make fire, yet there is still no completely convincing scientific explanation. For some time, it was believed that moths mistake a bright light for the moon, and attempt to navigate by maintaining a constant angle to it. Whereas the moon is so far away that its position relative to the moth changes only infinitesimally however far the moth flies, similar behaviour towards a closer light source would result in the moth circling it. However, it would not necessarily lead to the moth approaching the light ever more closely, which is what usually happens. A different if perverse explanation holds that the moth is dazzled by the light, and is really trying to avoid it by flying into the zone of darkness that, because of the way the moth's compound eye works, seems to surround it. Young (1997) goes into more detail on this subject.

Moths certainly do behave as if dazzled by a very bright light, flying erratically and bumping into solid objects. However, they are also attracted to the gentler illumination of a lighted window, less intense than daylight and

Plate 84: Although an abundant and ubiquitous moth, the Yellow Shell *Camptogramma bilineata* is not often attracted to light. The female illustrated was from a small (inbred?) colony in a Brighton garden, most being much more strongly marked than those on nearby downland. (Sussex).

therefore well within the capacity of the moth's eye to adjust. Here, their behaviour suggests that dazzling is not an obvious factor. Possibly more than one response is involved, depending on the wavelength and intensity of the light. Although it has often been shown that the brighter the light source, the more moths it attracts (both in terms of individuals and variety), some species are exceptions, and are more likely to be seen fluttering at the kitchen window than in the moth trap only a few metres away. These include geometrids such as the Silver-ground Carpet and the Yellow Shell (Plate 84) in the subfamily Larentiinae – but not those in the Ennominae, on which ultraviolet light has an especially powerful effect. Certainly any explanation of why moths are attracted to light must take into account its unequal effect on different, even closely related, species. It should also explain why insects other than moths are attracted, especially craneflies, caddis flies and some beetles. Birds are also attracted to light, in certain circumstances, notoriously to lighthouses and oil-rig gas flares when on migration in conditions of poor visibility. In easterly haar, my own powerful moth trap has brought down migrating Redwings and Fieldfares in autumn. More alarmingly, one foggy night a flock of Greylag Geese circled it noisily at treetop height for several minutes.

Improvised moth traps

However unsatisfactory the scientific explanation for moths' susceptibility to light might be, taking advantage of this attraction is easy. Leaving the light on all night in an upstairs bathroom with the window open can be quite productive, if there is no objection to the numerous other flying insects that will also enter. (Bathrooms offer fewer places for moths to hide than furnished rooms.) A lighted porch will function very well as a moth trap, especially if a stronger bulb than would otherwise be necessary is used. However, checks are best made during the night, as most moths will fly away at dawn. Outside lights on a house wall, especially if fixed fairly high, will again attract moths. These are useful early and late in the year, when it is not worthwhile running a proper trap. All the above techniques work better in the countryside, where there are no other nearby lights. In towns, there are so many street lamps and lighted windows that they tend to cancel each other out, diminishing the pull of any single one. There is also the intriguing possibility that urban moths, through the process of natural selection, are already becoming less susceptible to such potentially fatal distractions from their normal activities.

Commercial moth traps

Fry and Waring (1996) discuss the merits, drawbacks and relative performances of the various types of light trap that are commercially available in Britain from specialist dealers, who advertise in the entomological journals.

The Heath Trap

This is the cheapest, though least powerful, design. It is based on a fluorescent tube producing actinic light with a high ultraviolet content, the most attractive wavelength to moths. The tube is set vertically, surrounded by three plastic vanes that intercept moths approaching the light and cause them to drop through a funnel into the holding compartment. This is a rectangular box made of thin sheet metal, designed to pack flat for easy transport. Power is supplied by a 12-volt battery. A car battery will serve, but it is better to purchase one designed for camping or use on small boats, incorporating a handle for easier carrying.

Although the trap itself is not heavy, the battery weighs about 14 kg. Thus the equipment is only relatively portable. It is much used for scientific research, where a series of standardised traps is required to sample the moth fauna in different habitats simultaneously. In the most remote areas, the trap is often left unattended for the whole period of the study, and fitted with a photoelectric cell to turn it on and off each night. Regrettably, a killing agent inside the trap is then necessary to prevent too many moths escaping, or becoming so worn as to be unidentifiable without dissection.

Otherwise, Heath traps are most often employed on mothing expeditions to supplement the larger and more powerful mercury vapour light traps that require mains electricity or a portable generator, which in practice limits the places where they can be sited. The catch in the Heath trap will be much lower, perhaps a tenth as many individuals of a quarter the number of species, with a bias towards geometrids. However, there are numerous anecdotes relating how all the really good, sought-after moths turned up in the Heath trap, whereas the larger one was filled to overflowing but only with the commonest species. Perhaps this is because the Heath trap can more readily be set right amongst the most suitable habitat, however awkward the terrain, but the different strength and wavelength of its light might also play a part.

Of course, there is no reason why a Heath trap should not be employed in a garden. If so, it can be adapted to run from the mains, which also increases its effectiveness. Being less conspicuous and obtrusive than a more powerful trap, in built-up areas it is not as likely to suffer from vandalism or incur the displeasure of neighbours. For the beginner, starting to learn and identify the different species, there are positive advantages in not catching an overwhelming number of moths at once.

The Robinson Trap

The study of macro-moths has been revolutionised since traps fitted with a powerful mercury vapour bulb, emitting ultraviolet light, came into widespread use. Named after the two brothers who first developed it, the Robinson trap became commercially available in the 1950s. It proved so effective that it is still the ultimate weapon in the armoury of the dedicated moth hunter. Inevitably, such a specialised item of equipment with a low sales volume is expensive. Adding on the cost of spare bulbs, and a portable generator if the trap is to be used more than the cable's length away from a

mains electricity supply, makes this a major purchase. It is definitely not an impulse buy. However, those who do succumb clearly have no regrets: second-hand ones are rarely advertised for sale.

Originally the body of the trap was metal, and suffered from condensation on cool, still nights. Modern versions, in rustproof plastics and Perspex, are cleverly designed and well made. The usual light source is a 125 W MB/U bulb or equivalent, which cannot be run from an ordinary domestic light socket as it requires a choke. This bulb emits light from the ultraviolet end of the spectrum. Although it appears intensely bright, only part of the output is visible to human eyes, whereas the eyes of moths are particularly sensitive to the wavelengths not registered by ours. As insects approach the bulb, baffles intercept their flight, usually causing them to drop through a funnel into the main body of the trap. This is loosely stacked with papier-mâché egg trays, which provide a rough surface giving a secure foothold, and many dark nooks and crannies where moths can hide. Most quickly settle down, perhaps assuming it is dawn.

The Skinner Trap

This trap, named like the others after its inventor, uses the same electrical equipment and power supply as the Robinson trap, but its body is box-like, packing flat for easy storage or transport. The hinged sides are made of sheet aluminium, and support a horizontal strut to carry the light fitting. Two sloping panes of Perspex or some other clear plastic form the top, and channel moths downwards until they fall through the 3 cm gap between them and into the interior of the trap. There is no base, so the trap is placed on an old bed sheet laid on level ground.

Besides being significantly cheaper than the Robinson trap and easier to transport, the Skinner trap has the advantage that moths can be more easily removed for examination while the trap is operating. Thus it is ideal for 'Moth Evenings', and other attended sessions. However, moths can escape from it more easily, both at night and after dawn, so a greater proportion of the catch is lost.

Safety Precautions

All electrical equipment intended for outdoor use must be sturdy and well maintained, connections waterproof and cables undamaged. This applies particularly to traps designed and built at home, using standard electrical parts, a much cheaper option for those with the requisite skills. It is also very advisable to fit a plug incorporating a circuit breaker in the event of a power leak. If the trap is run in a garden, note that the householder is responsible for the safety not only of all legitimate callers, but even of trespassers and felons to some extent.

Concern is sometimes expressed about potential eye damage from both the invisible ultraviolet wavelengths and the very bright light from the powerful bulbs used in moth traps, though it seems that none has ever been reported. A special coating inside the bulb transmutes the most dangerous wavelengths, which is why a bulb with a broken outer casing should be discarded even if it still works. Nevertheless, Crafer (1998) advises the use of polaroid sunglasses when working a moth trap at night. However, Bailey (1998) points out that even ordinary spectacles cut out 90 per cent of the ultraviolet light, though not the dazzle. It is certainly best to avoid staring directly at the bulb when working the moth trap, and regulars soon develop the automatic habit of shielding the eyes with an arm.

As with other forms of mothing, it is always safer (and more fun) to have company when light trapping at remote sites after dark. Arrive in daylight and check the place out; if there are any disreputable characters around, go elsewhere. A portable generator is a tempting target for thieves. Normally, though, the greatest danger is that of being eaten alive by midges on a warm still night. In compensation, such nights are usually good for moths too.

Public relations

A moth trap is almost certain to arouse curiosity, and rightly so. Neighbours, and members of the public, are entitled to know what is going on, and why. Far from being a chore, this duty to explain should be seen as an opportunity to promote an interest in moths, their habitat, and wildlife in general. Some people are good at public relations, able to convey their enthusiasm in a relaxed, easy manner. They naturally assume the role of spokesperson in a group. Others like myself are forced to work harder.

In towns, running a powerful moth trap is a potential source of friction with neighbours. A laudable concern for the welfare of the moths is easily dealt with, but a valid objection to the powerful light shining all night is a different matter, especially if there is worry about the ultraviolet content. It might be necessary to arrive at a compromise, and only run the trap on certain nights, or turn it off at midnight. This will of course reduce the catch, perhaps a worthwhile price to pay to maintain harmonious relations. Feuds between neighbours can be bitter. Or it could be possible to screen the trap from certain directions by erecting low panel fencing. In my own rural garden, with the nearest neighbours over half a kilometre away, I use such screens to avoid dazzling passing motorists. They also provide shelter on breezy nights, and somewhere for moths to sit, outweighing any slight reduction caused to the catching radius of the trap.

Aiming to forestall any objections from neighbours, when introducing a Robinson trap to a Brighton housing estate I fitted it with a 'black bulb' having an outer envelope of Woods glass. This allows through ultraviolet wavelengths, but cuts out virtually all visible light. Only an eerie purplish pink glow could be seen when the trap was switched on. Unfortunately, many of the dyes in the curtains, upholstery and wallpapers of the surrounding houses fluoresced alarmingly in ultraviolet light. The effect was truly weird. Predictably, when I put the trap out the following evening, my nosey next-door neighbour was soon leaning half out of his bedroom window observing me. Since we did not get on, I pretended not to notice him. When his curiosity finally got the better of him, and the inevitable question came, there had been ample time to think of a smart answer. It was, I told him, a device I'd invented for signalling to flying saucers. This was the autumn when crop circles first hit the headlines, and my facetious reply was all too readily believed. It was just as well that we had already arranged to move house, and left the estate shortly afterwards.

Care of the catch

On a good night at the height of the summer, a Robinson trap often attracts hundreds if not thousands of moths. The trap is not selective, so the great majority of the moths will be unwanted except as a source of data, but they will be caught anyway. Most trappers feel that they have some duty of care towards their catch, and want to do all they can to minimise casualties and reduce as far as possible any disruptive effect both on the individual moths and their populations.

Often, moths that approach a light trap do not enter it but land short, where they are in danger of being trodden on, especially at group sessions where many people are gathered round. Placing the trap on a white sheet helps to avoid this. It is not so necessary in a garden where a trap is run regularly; here the trap is usually put on a tightly mown lawn. The grass should not be cut if the trap was operated the previous night, as moths like the Large Yellow Underwing are adept at hiding in the turf.

Domestic cats are a nuisance in some gardens, even dismantling a trap when trying to pounce on the moths fluttering inside. Sticky tape will help to hold it together. Nothing can be done about the bats that, at regularly used sites, soon take advantage of the moths and other insects drawn to the light. They would only be catching moths elsewhere in any case. Frogs and toads sit and wait for the occasional straggler, and can be tolerated too. Birds are a different matter. They are quick to discover that the moths outside the trap provide an easy meal at dawn, for themselves or their importunate fledglings. Worse, the more intrepid sometimes learn to enter the trap itself and wreak havoc among its contents. Wrens, House Sparrows and Great Tits, at home in confined spaces, are the most usual villains. The only solution is to rise before the lark, collect up the more visible moths on the ground, and pop them in the trap. As a reward there is sometimes a really exciting find, which would have flown away as the day brightened even if it had not been eaten. I well remember my first Brussels Lace, sitting on top of the Perspex lid, providing a new vice-county record too. The trap can now be brought indoors, its funnel blocked with crumpled tissues (or old socks) to prevent moths escaping. Then, go back to bed!

After the night's catch is examined and counted, there is the question of the best method of release. Hopefully, the days are long gone when unwanted moths were simply tipped out on the lawn to provide a feast for waiting birds. Present guidelines recommend that the trap is kept all day in a cool shady place, and emptied at dusk, although any moths that do not settle down should be allowed to leave. Some species are partly diurnal anyway. However, it is also important that moths are released at or near the site where they were caught, and not transported to different areas or habitats that might be unsuitable for them. Sometimes, for instance on field trips, this means that the trap must be emptied in the daytime. If so, the moths should be scattered in dense vegetation such as heather or long grass where they can hide.

Even in a British summer, care should be taken that the trap and its contents do not overheat, with fatal results. It is easy to forget how far the sun (or rather, the earth) moves: a spot that was in deep shadow during early morning might be exposed to the sun's full glare a few hours later. Dehydration is a serious threat, especially in very hot weather, as moths in the trap have no opportunity to feed. Placing a nectar source such as a spray of Buddleia inside the trap remedies this. A wad of paper kitchen towel soaked in a weak sugar solution will also serve, and only takes a minute to provide.

Inevitably, if moths are released at dusk where they were caught, and the trap is operated again that same night, a sizeable proportion will be recaptured almost immediately. Where a trap is operated continuously, some moths must spend most of their lives inside it. Although unfortunate for the individuals concerned, there is no evidence that it has any effect at the population level, even for the scarcer species. On the other hand, proof either way would be almost impossible to obtain, unless the effect was catastrophic. Perhaps it helps that most of the moths caught in light traps are males rather than the all-important females.

A further consideration is the effect that a high percentage of retraps will have on the data. Unlike bird ringers, many moth trappers seem unaware of the extent of this problem, adding together the total for each session as if each capture was a new individual. Whilst it is possible to mark moths for a special study, it is quite impractical to do this routinely. However, a few can be recognised again from distinctive markings, slight deformities or scars, showing that the same moth may be caught several times – not necessarily on consecutive nights – over a period of a week or more.

For both humanitarian and scientific reasons, therefore, some trappers choose not to operate every night even if the weather is suitable. Especially when sated after a bumper catch, they prefer to give the moths (and themselves) a rest. When working a new site, or having acquired a trap for the first time, it is naturally tempting to use it as often as possible. However, experience teaches that equally good results can be obtained with much less effort by being selective, and only putting the trap out when conditions are promising. If nevertheless the trap is to be run on consecutive nights, perhaps because scarce migrants are known to be arriving, it may be possible to release the previous night's catch in similar habitat nearby. At my site, moths released 100 metres away are commonly recaught, but this happens much less often when the distance is increased to 200 metres. Another solution is to run the trap only for the first half of the night. Unless conditions are ideal, this is normally the most productive period anyway. Moths can then be examined and released immediately, leaving them the rest of the night to disperse.

In practice, the British weather makes it impossible, or at least pointless, to run a moth trap on a sizeable proportion of nights – it might be too cool, too clear, too windy or too wet. Temperature is by far and away the most important factor: the warmer the night, the more active moths will be. It is mainly when the temperature is on the low side (under 10°C) that the other factors come into play. Geometrids in particular do not like wind, but will fly in surprisingly cool conditions if it is calm. Spring and late autumn moths are necessarily more tolerant of low temperatures than summer ones, but even for these 5°C can be regarded as the minimum. However, predicting whether it will be a good night for moths is more of an art than a science. Partly this is because the night must be considered in the context of the previous ones. If moths have been forced to sit out a long spell of poor weather, they will fly once conditions improve even moderately. However, an identical night that represents a deterioration after a spell of good weather is likely to be poor.

Surprisingly, showery nights are often excellent for moths, especially the more powerful fliers like noctuids, as the clouds help to keep the temperature up. Unfortunately, rain and electrical equipment do not go well together. Mercury vapour bulbs run very hot, and can be shattered by cold heavy rain. The flimsy shield supplied for the Robinson trap is effective only against light, vertical drops. To protect the bulb against heavier, angled downpours, an inverted Pyrex glass bowl is often placed over it. I consider that this practice causes unacceptable injury to moths. A pocket of heated air is trapped under the bowl, and any moth that flies into it is badly harmed. The most obvious damage is to the antennae, which are frizzled in the heat, and the eyes too are affected, leaving the moth disabled. Far better to forego a trapping session if heavy rain is forecast than inflict such casualties.

The effectiveness of light traps – a mixed blessing?

Operating a modern mercury vapour light trap is by far the quickest, easiest and most successful way of obtaining moths. It does not work for all species: for obvious reasons, day-flying ones are hardly ever attracted – perhaps only when disturbed from nearby resting places. Those that fly mainly at twilight are also greatly under-represented. Relatively few microlepidoptera are truly nocturnal, and even those that are tend not to be strongly attracted to light, or to enter the trap itself if they approach it. (An amusing exception is the enterprising White-shouldered House-moth *Endrosis sarcitrella*: a colony has thrived in my Robinson trap for many years, feeding on dead flies and other debris.) However, for luring macro-moths in particular, it cannot be denied that the light trap is supreme.

Using a wide variety of methods at my home site in Banffshire, I recorded 261 species of macro-moth during the 10-year period 1990–99. (This would be an unexceptional 'garden' total further south; here it represented 84 per cent of the vice-county list.) Of these, no less than 40 were seen only in the Robinson trap. They were not found by any of the other techniques, either as adults or caterpillars. However, 18 of them were recorded just once in the 10 years, strongly suggesting that they were simply migrants or strays. Of the remainder, only 6 were considered probably resident and a further 6 possibly so, the rest almost certainly being short-distance strays too.

So it might be assumed that, without a Robinson trap, my site list would have reached only 221 species. This does not necessarily follow, since without the luxury of the trap I would have put more effort into other types of fieldwork. This might well have produced sightings of some of the extra 40 species, especially the probable residents, but less likely the 'once in a decade' strays. Nevertheless, there is no doubt that without a Robinson trap my list would have been shorter. But, in a sense, it would have been more accurate – a more faithful picture of the moths that bred in the area, or were regular visitors. It would have contained fewer wanderers from the coast 11 km away, like the Marbled Coronet and The Confused *Apamea furva*. Most of the other obvious strays (moths restricted to food plants that do not occur on the site), would also be missing, like the Poplar Lutestring *Tethea or* (Aspen), Juniper Pug *Eupithecia pusillata* and Orange Sallow (lime). As two or three of these casual species are still being added annually, it could be argued that my site list is becoming more misleading with every year. This is mainly a criticism of the simplistic 'presence or absence' method of recording that takes no account of status. The light trap itself provides clear and accurate data: the problem is one of evaluation and presentation. Lists should be annotated, dot maps should reflect numbers and regularity of sightings. Even then, they need to be interpreted with insight and caution.

Nor should the occurrence of vagrants or strays at a site, and the light trap's role in revealing them, be dismissed as trivial. It is valuable to know that some individuals of species normally regarded as sedentary do disperse, or allow themselves to be blown by the wind away from their usual breeding areas. (Moths normally avoid flying in very windy conditions.) This gives us confidence that the present breeding range and distribution of most moths accurately reflects their requirements – they are not absent from suitable areas

Plate 85: Scarce Prominent *Odontosia carmelita*. Before the invention of mercury vapour light traps this much-coveted species was very difficult to obtain either as an adult or in its early stages. (Banffshire).

because they cannot reach them. It shows that they have the scope to colonise new ground if conditions change to suit them. Very few strays, of course, will successfully establish new populations. If this were the aim, the behaviour would be confined to mated females, but males stray too. Probably, genetic mixing is the main function, or at least the beneficial result, of straying.

A more valid objection to the sheer effectiveness of light trapping is its tendency to make other methods of finding macro-moths redundant. Fieldwork skills, no longer needed, soon wither. In many ways, the Victorians knew more about the ecology of moths than we do, though they would not have thought of it in those terms. Take the Scarce Prominent *Odontosia carmelita* (Plate 85), a local, low density and very desirable species. Tutt's (1901–05) compilation of practical hints contains no less than 13 (somewhat repetitive) entries on how to obtain this prize. Who today would root around at the base of birch trees in autumn for its pupa, or search their trunks at midday in April to find newly emerged adults? Just switch on the mercury vapour light trap!

The Alder Moth is another former rarity now easily obtained by the same method. Once, it was hardly ever found except as a caterpillar 'and anyone who may obtain even a single example in a season may congratulate himself on a good find' (South, 1907–09). Yes, but what a thrill that must have been!

Operating a light trap is exciting too. There is no other way of seeing moths in such numbers and variety, including species hardly ever found by other methods, perhaps because they hide so well by day, or do not feed and therefore never come to sugar or natural attractions (Plate 86). Often, in practice the choice is to see a species at mercury vapour light, or not see it at all. Equally, using light traps to the exclusion of all other methods removes much potential enjoyment: the satisfaction that comes from successful fieldcraft, and the pleasure of observing moths behaving naturally in their proper environment. Invaluable tool, or lazy short-cut: the light trap can be both, depending on the circumstances and the use made of the results.

Plate 86: Canary-shouldered Thorn *Ennomos alniaria*. Moths in this group do not feed, but are particularly susceptible to mercury vapour light. Modern traps have shown them to be commoner than once thought. (Banffshire).

Chapter 8

Searching for the early stages

The two previous chapters have explored the different ways of finding or attracting moths. However, for various reasons, some species are rarely seen as adults. This may be because of their behaviour: perhaps their habits are not very well known, or make them hard to catch. Light might not attract them strongly, for instance if they are crepuscular. They could be diurnal, but spend their time high in the treetops out of reach, like some clearwings. Or they might hide in thick cover during the day, and refuse to take wing if disturbed. Whole families of moths do not feed as adults and are not interested in nectar sources or sugar. Such species are often easier to find in their early stages as eggs, caterpillars and pupae. Others, especially some of the microlepidoptera, are more easily identified then too, because of the distinctive mines, spinnings and cases the caterpillars construct on their particular food plant. An expert can reliably record them from a site on this evidence alone, even if the habitation is empty! Finally, because females lay many eggs but few complete their metamorphosis, adult moths are always outnumbered by their early stages, sometimes dramatically so. In Sussex, the silk webs of Brown-tail caterpillars were a familiar sight on defoliated Hawthorns, but in some years I never saw a single adult moth.

Whether or not these factors apply, the early stages of moths are worth studying in their own right. Caterpillars in particular provide equally striking examples of camouflage, or of warning coloration, as adult moths, and their behaviour can be just as interesting. Life histories are very varied, and the successful unravelling of them helps to answer many otherwise puzzling questions about distribution and abundance. Family relationships are sometimes more obvious in the caterpillar or pupal stages, shedding light on the system of classification. And is it truly possible to know a species without having seen its caterpillar in the wild, or even bred it in captivity?

Finding eggs

Moths lay numerous eggs, but they are small. Often, they are very well hidden, pushed into cracks in the bark, or inside sheaths of grass stems. Finding a needle in a haystack would be easier. Sometimes, a female can be observed in the act of laying, but this is less easy with nocturnal species. More often eggs are found by chance, and then identification can be a problem, even if the food plant offers a clue. It might be necessary to wait until the eggs hatch and the caterpillars are well grown before their identity can be established.

Plate 87: Egg batches of the Kentish Glory *Endromis versicolora* are easily recognisable, often being laid in a double row on the outer twigs of young birch saplings. (Moray).

Deliberate searching for eggs with any reasonable hope of success is practical for only a few species. The brown hemispheres of the Puss Moth can be found, usually in pairs, on the upper surfaces of poplar or sallow leaves. The rounder, pale green eggs of the Poplar Hawk-moth might be seen at the same time. The very large batches laid by the Large Yellow Underwing, each egg meticulously placed, also attract notice in late summer, as do the bead-like eggs of the Vapourer Moth on the female's cocoon. Looking for eggs in the wild, however, is the favoured way of obtaining or recording very few species. The Kentish Glory is perhaps the main exception, as its large purplish-brown eggs are laid in very distinctive batches at a convenient height near the tips of birch twigs in spring (Plate 87). With practice, they are easily found at the moth's remaining sites before the leaves open fully and hide them.

Finding caterpillars

Almost everyone knows what a typical caterpillar looks like, an education process begun by children's books and reinforced by personal experience. A caterpillar, it might be thought, is unmistakable – impossible to confuse with anything else. Yet it is surprising how often sawfly grubs are mistaken for caterpillars. Having the same lifestyle, feeding openly on the leaves of plants, they have evolved a similar structure and appearance though belonging to a different order of insects, the Hymenoptera. They are a fine example of convergence. As a child I spent many hours poring through my moth books, trying without success to identify the bright orange, black-spotted 'caterpillar' that was common on the local poplars. No wonder it was not in my books: it was the larva of a sawfly. Even today, I sometimes have to look carefully when faced with an unfamiliar species. Fortunately, there is a simple check. Sawfly grubs have a pair of legs on almost every segment, but in caterpillars there is always a gap between the thoracic legs and prolegs, with at least two and up to five of the intervening abdominal segments having no legs at all.

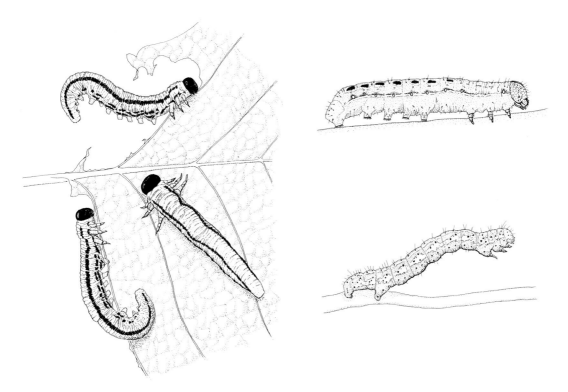

Figure 6: Sawfly grubs (left), with two caterpillars (right) for comparison.

(This does not apply to the minute early instars of some leaf-mining microlepidoptera, when legs may be absent altogether.) Many sawfly grubs also hold their abdomen in a characteristic curled position (Figure 6).

Caterpillars are often highly cryptic, or good at concealing themselves. Thankfully, their habits, when understood, offer some help to the determined searcher. While feeding and growing, they must remain within reach of their food at all times, and cannot risk starvation by wandering too far away. This concentrates the search. If a species is confined to a single food plant, or restricted to a small range of these, this naturally helps with the targeting of effort. Indeed, it is generally far easier to find such specialists on demand than it is to find polyphagous caterpillars, where the search must inevitably be broader. But nearly all caterpillars have an Achilles heel: since they are too small to consume the entire plant in a single meal, the damage caused by their feeding creates tell-tale signs that predators, and lepidopterists, are able to exploit. Only a few internal feeders are able to complete their growth giving no visible indication of their presence. Most of the following methods of finding caterpillars rely on the clues they leave.

Visual searching

This is probably the most skilled, and therefore the most satisfying, way of finding caterpillars, though not necessarily the most productive. If a particular species is the target, the obvious course is to read up its habits in the literature, then choose the correct time of year to work the appropriate habitat and food plant. However, much searching for caterpillars is opportunistic, prompted by noticing evidence of recent feeding. Usually the damage is to the leaves of a plant, but some species prefer the flowers and seedpods, or bore into stems and roots, in which case frass (pellets of droppings) and entrance or exit holes provide the clues.

Plate 88: When too large to resemble the needles, one form of the Pine Hawk-moth *Hyloicus pinastri* caterpillar mimics a whole spray, its brown dorsal stripe representing the knobbly twig. (Norfolk).

Various other invertebrates eat plants too, so first it must be established that caterpillars are the likeliest culprits. This is not always easy. Slugs and snails feed by rasping, thus the holes they make in leaves are ragged and rough; often, there are trails of slime too. The grubs of leaf-eating beetles usually feed communally, skeletonising leaves with many small holes. Caterpillars make clean cuts with their sharp mandibles, methodically eating neat portions of a leaf at one sitting, at least when larger. However, the damage caused by sawfly grubs is often indistinguishable (to me at least). Leaf-mining insects include many flies as well as microlepidoptera. This can also cause confusion, but in moths the frass inside the mine tends to be more carefully deposited in lines, rather than scattered randomly. Inevitably, the lepidopterist searching for caterpillars makes the acquaintance of other invertebrates too.

Caterpillar damage to leaves may be several months old. Often, the individual that caused it has long since gone. Close scrutiny will show whether the bitten edges are moist and therefore recent, as within hours they become brown and sealed to prevent loss of sap. The presence of frass is another favourable sign because it soon disappears, disintegrating in rain or perhaps being eaten by slugs. Normally frass falls to the ground where it is hard to spot, except on town pavements, but sometimes it lodges on the lower leaves of the plant. When caterpillars are very numerous this shower is actually audible. Several times in southern oakwoods the pattering sound has fooled me that it was starting to rain in spite of a cloudless blue sky. It may have been imagination, but listening carefully I could hear hundreds of tiny jaws in action.

Many caterpillars construct a shelter for themselves amongst the leaves of their food plant. Sometimes this is no more than a few flimsy strands of silk holding leaves loosely together, like that of the Winter Moth, easily torn open by birds such as Blue Tits. Others are stronger and more elaborate, and characteristic of a species or group. Some tortricids and pyralids neatly fold

Plate 89: The Dark Tussock *Dicallomera fascelina* caterpillar, although effectively protected from Meadow Pipits by irritant hairs, is surprisingly well camouflaged in its heathery haunts. Since hairy caterpillars are eaten by Cuckoos (frequent on moorlands) this dual strategy is presumably the safest option. (Banffshire).

leaves over, or roll them into tight cylinders. The Poplar Lutestring uses small dense pads of silk to spot weld two Aspen leaves together and form a pouch. Numerous species, both micros and macros, spin together the terminal leaves of a growing shoot. It is thought that this also depresses the tannin content of the young leaves, making them more palatable. Certainly, most caterpillars feed on the leaves they have spun together, for example the Argent and Sable. Others, like the Ruddy Highflyer *Hydriomena ruberata*, carefully avoid eating their shelter and use it purely as a daytime bivouac, coming out at night to feed elsewhere on the plant.

Either way, the dwellings are relatively conspicuous: much more obvious than a cryptically coloured caterpillar itself would be. Nor are even the most well-constructed spinnings impregnable, but presumably they do alter the economics of foraging. It is hard for birds to perch on the tips of slender shoots; then there is the time and energy it takes to open up the spinning. Many will be empty: perhaps on average only a third contain a caterpillar which may prove to be disappointingly small. Moreover, some caterpillars will escape (those of micros are adept at running backwards) and drop to the ground. When there is easier alternative food, it may not pay a bird to investigate spinnings. Fiddly to open, with no guarantee that the contents are worth eating – anyone who has struggled with modern tamper-proof food packaging will sympathise!

Having noticed the circumstantial evidence, the next step is to find the caterpillar itself. The distasteful species present least difficulty, as they make little effort to conceal themselves, being protected by tufts of hair, bright warning colours, or both. Their survival depends on being clearly seen and instantly recognised by would-be predators. Eggars (Plate 45), tussocks (Plate 89), tigers, and noctuids of the genus *Acronicta* (Plates 40, 41) feed or rest openly by day, and consequently figure large in many a naturalist's childhood memories, including my own. The caterpillar of the Garden Tiger, resplendent with dark silky hairs, was then a great prize, kept almost like a pet. According to the books it was known as the Woolly Bear, but the boys from the rougher part of town called it a Hairy Mary. My parents forbade me to copy them, but refused to explain why. It was many years before I understood the allusion.

Heaths and moorland have an unusually high percentage of these hairy caterpillars. Perhaps the small leaves of heathers offer scant concealment for larger diurnal species, making unpalatability the more successful strategy. Equally conspicuous are the bold black and orange caterpillars of the Cinnabar (Plate 53), busily devastating coastal ragwort, and the polka-dotted ones of the Mullein Shark (Plate 90), doing the same to the flowers of its namesake. Splendid though it is to see such attractive and colourful caterpillars, finding them can hardly be regarded as a feat of detection. It is the cryptic species that provide the real challenge. Many of these feed and rest openly too, relying entirely on their shape and colouring to ensure that they are seen yet not seen, or rather not recognised for the tempting packets of protein that they are. They provide wonderful examples of camouflage, hardly surprising since their lives depend on it. Outwitting designs that have been perfected over millions of years can give great enjoyment and satisfaction.

My own favourites include the prominents, as varied and attractive as the adults. Some have camel-like humps and unusual resting postures, as well as disruptive markings and

Plate 90: Unlike most aposematic caterpillars, the Mullein Shark *Shargacucullia verbasci* is not hairy. Its warning colours are effective against birds, but not against parasitic flies and wasps, major enemies of this and related species. Caterpillars that survive give rise to highly cryptic adults (Plate 55). (Sussex).

Plate 91: Coxcomb Prominent *Ptilodon capucina* caterpillar. This unusual posture is characteristic of the species, presenting the red thoracic legs and twin tubercles near the tail as if they were defensive weapons. Possibly they do secrete substances to deter predatory invertebrates, but this has never been studied. (Banffshire).

counter shading to break up the familiar caterpillar outline (Plates 11, 39, 91). This is highly effective amongst dappled foliage at close range, and is widely employed by other families as well. Curiously, such caterpillars are often easier to spot from a distance. As a boy, I learnt to find those of the Eyed Hawk-moth on sallows growing in the bed of a drained canal in Lancashire by standing 6 or 7 metres back, as this negated the shape-destroying effect of the pale diagonal slashes down their sides. Even then, without keeping my eyes fixed on the caterpillar during the approach, I often could

not see it after reaching the bush.

A different technique is needed for the 'stick caterpillars' in the family Geometridae. Only the closest scrutiny can distinguish them from the twigs on which they sit, rigid and immobile. So good is their resemblance that often it is necessary to touch one – or a twig – to make certain which is which. Ironically, its very perfection sometimes gives it away, for the copy arouses suspicion by being too convincing (Plate 92).

Plate 92: The Pale Brindled Beauty *Apocheima pilosaria* caterpillar is typical of many in its subfamily that resemble twigs. Because they feed on almost any native deciduous tree, the match cannot be precise: this individual on birch was noticed through being more twig-like than the twig itself. (Banffshire).

Soil sifting

Sharp eyes alone are insufficient to find those caterpillars that feed only at night, and leave the food plant to hide during the day. Many noctuids stay on the plant when small, but adopt this strategy in their later instars. Like the adults they will become, they tend to be cryptically coloured in a general-purpose way, rather than tailored to a particular food plant or situation. Thus they are usually a dull shade of brown or green, with various understated paler and darker stripes, lines, dots, dashes and chevrons to disguise their shape. So many species adopt a similar drab livery that identification can be a problem, especially as both ground colour and strength of markings can vary between individuals, and change as the caterpillar grows and moults. Descriptions in the literature are normally of the final instar, but the earlier ones may look very different. Before Porter (1997) no comprehensive illustrated guide to the caterpillars of British moths was available in print.

Collectively, the noctuids are responsible for much of the damage to tender young plants that infuriates the gardener, who knows them as cutworms, or frustrates the lepidopterist because by morning there is no sign of the culprits. Like naughty schoolboys, they have distanced themselves from the evidence of their misdeeds. They are lying low, under withered basal leaves, in the litter and detritus below the plant, sometimes just beneath the surface of the soil. Those that climb trees in the spring to feed on the new growth often spend the day behind loose bark near the base of the trunk, or in the leaf litter.

One way to find these nocturnal feeders in the daytime is to rootle around like a Blackbird near the base of the plant, turning over dead leaves and raking through the surface debris. Sometimes a minor mystery will be solved. For several years I searched the Sea Campion growing on the Banffshire coast, hoping to add Marbled Coronet to the vice-county list. Although the contents of almost every seed capsule had been plundered in the characteristic way, there was never a caterpillar to be seen. Finally, I tried digging in the soil below the plants for pupae, and discovered this was where the caterpillars had been hiding all along. The habit was not mentioned in any of my books, but that is not unusual or necessarily to be regretted: finding out such things for oneself gives greater pleasure. Sieving the sand of vegetated dunes is the recognised way of finding caterpillars of coastal noctuids such as the Sand Dart and Portland Moth.

Nocturnal forays

An obvious alternative approach is to look for nocturnal caterpillars when they are active at night, with the aid of a torch. This can easily be combined with sugaring or working natural attractions for adult moths. It is particularly effective along woodland edges and rides on mild nights in early spring, when many species climb shrubs and saplings at dusk to feed on the bursting buds, and there is not yet sufficient foliage to hide them. The external feeders on grasses are also readily found by torchlight, showing up well in the beam. However, too strong a torch, or a clumsy approach, will cause them to release their hold and drop out of sight.

Beating and sweeping

These are effective, if indiscriminate, ways of obtaining caterpillars (and not only those of carpet moths). In beating, an old sheet, inverted umbrella or purpose-made beating tray is placed directly below a branch of a tree or shrub, which is then struck with a stout stick. The aim is to dislodge any caterpillars that might be present, causing them to fall onto the sheet or tray where they are easily seen. If done repeatedly and too vigorously, beating damages the vegetation, hence it is frowned upon by some nature reserve managers and naturalists. Such violence is unnecessary, since one sudden unexpected jar is quite sufficient to cause caterpillars to lose their hold, if they are going to do so, and subsequent thrashing merely causes them to cling more tightly. The same technique can be employed on a smaller scale with tall herbaceous plants or dwarf shrubs, using a biscuit tin lid or a butterfly net to intercept the falling caterpillars.

Plate 93: Vast numbers of caterpillars exploit the young foliage of oaks in spring, before tannin levels rise. The Spring Usher *Agriopis leucophaearia* is one, and typical of many in its matching leaf-green uniform, stretched out along a main rib. Such caterpillars form part of the mixed bag readily obtained by beating. (Norfolk).

Plate 94: Feeding mainly on heather, the caterpillar of the Grey Mountain Carpet *Entephria caesiata* can tailor its camouflage exactly to match its food plant. Because feeding damage to heather is not obvious, sweeping is the most efficient way to obtain this species and others that share its habitat. (Banffshire).

Species that spin shelters for themselves are rarely dislodged by beating, except at night if they come out to feed. Otherwise, the bag is a mixed one, with numerous other invertebrates (including occasional moths) and assorted debris besides caterpillars. As with light trapping, often only a small proportion of the catch is wanted, so the many are inconvenienced for the sake of the few. Although those caterpillars not required may simply be released, perhaps not all will regain the safety of the tree.

Sweeping is productive in low ground cover such as grasses, heather and bilberry, or arable weeds around the edges of fields. As with beating, better results are sometimes obtained after dark. The net is made of strong material, resistant to abrasion and tears. It is swept through the vegetation capturing an unknown proportion of the caterpillars present as well as various other creatures, seeds, and bits of plants. There are disadvantages: inevitably, some of the caterpillars are injured in the process. In mixed plant communities, it is not always possible to determine which was the food plant, let alone what part of the plant was being eaten, so the value of the data is reduced. Nevertheless, it can be an effective way to acquire caterpillars that are not easily found by searching (Plate 94).

Luring

It may come as a surprise that caterpillars can be lured. True, the occasional one is seen on the sugar patch, but this is hardly a regular event. However, especially in fenland, caterpillars can be attracted to litter heaps created by cutting herbage and arranging it into piles roughly 40 cm high and about 1 metre in diameter. After a few days, the heap is carefully picked up and shaken in a sieve. Any caterpillars hiding in it fall through the mesh and onto a sheet spread below. Although a wide variety of species that feed on grasses or low plants may be obtained in this way, the technique is chiefly valuable for monitoring the Marsh Moth *Athetis pallustris* (Plate 95). The obscure and retiring habits of this scarce and local moth make it difficult to record by any other method (GM Haggett, personal communication). On southern heaths, sliced runner beans have attracted caterpillars of the Shoulder-striped Clover *Heliothis maritima*. Luring can work on a smaller scale too. When caterpillars of the Neglected Rustic *Xestia castanea* escaped and went to ground in my hopelessly cluttered study, I retrieved them all over the next few nights from strategically placed tufts of cut heather.

Pick and hope

Small caterpillars that live and feed entirely concealed within flowers and seed capsules are very hard to find by searching, for often there is no outward sign of their presence. In a known locality, it can be worthwhile to collect a quantity (be sparing!) of the food plant at random, hoping that some of it will be tenanted by the species in question. At home, it can be spread out on sheets of paper. Piles of frass will later reveal the whereabouts of any caterpillars in the sample. Species frequently obtained in this way include Toadflax Pug *Eupithecia linariata* (Plate 96), Slender Pug *E. tenuiata*, Barred Rivulet *Perizoma bifaciata* and many micros.

Chance finds

Whatever hunting technique is used, even the experts will have blank days, when all their effort and fieldcraft counts for naught. Fortunately, these are counterbalanced by chance finds that require no more than sharp eyes and the luck of being in the right place. Often these involve caterpillars in a hurry, urgently seeking a suitable place to pupate. It is the rapid movement that catches the eye. Tree-lined city streets provide many such easy pickings, for when a fully grown caterpillar descends the

Plate 95: Cut herbage piled in heaps at its fenland home will lure the dumpy, sluggish caterpillar of the scarce and local Marsh Moth *Athetis pallustris*; otherwise it is rarely seen. (Lincolnshire).

Plate 96: The Toadflax Pug *Eupithecia liniariata*, one of the prettiest of a large genus, can be bred by collecting bundles of its food plant speculatively. Sometimes it feeds on related plants in gardens. (Sussex).

trunk it meets, not soft earth, but impenetrable tarmac or concrete. Some interesting species can be found in this way. Nearly all the Lime Hawk-moth caterpillars I have seen were wandering along the pavements of central Brighton below the elms where they had fed. Others lose their foothold in gales. Once, what seemed to be a bright scrap of orange peel thrown in the gutter started to move: a Sycamore Moth *Acronicta aceris* caterpillar, with a hairstyle the envy of any punk. It seems only fair to rescue such venturers before they are crushed by heedless passing feet or tyres, and minister to their needs. This, no doubt, is how many observers begin to rear moths.

Finding pupae

In moths, this stage is usually well concealed, more so than in butterflies. The pupa is fully exposed to view in very few species, normally being encased in a cocoon, which itself is nearly always carefully hidden or camouflaged. Common sites are in the soil or surface detritus, behind dead bark, or on the food plant in a boring, spinning or folded leaf. Thus, for most moths, the pupal stage is not an easy one to find. Nor are pupae easy to identify, in spite of using the clues afforded by food plant, locality and time of year. Those of related species tend to be very similar, lacking any foolproof diagnostic features. The family to which it belongs is usually clear enough, but the species is often in doubt until the moth emerges.

Plate 97: Most cocoons of the Puss Moth *Cerura vinula* are so well camouflaged that they escape notice until breached by the emerging moth, or by a predator. (Sussex).

However, there are exceptions, where the cocoon is so distinctive that the moth can confidently be recorded on that evidence alone. On heathland, the smooth oval capsule of the Northern Eggar, and the unique exit-only construction of the Emperor Moth, are characteristic finds. Puss Moth cocoons, low down on trees or fenceposts, are extremely difficult to spot when first constructed, but become obvious once the moth has emerged (Plate 97) or Great Tits have hacked them open. They are so tough that their remains persist for years. Most familiar of all are the papery yellow tapered cocoons of the Six-spot Burnet, high on a waving grass stem. A similar but much larger version, blunt at one end, belongs to the Drinker Moth.

In the past, hunting for pupae was an important activity for collectors, especially in winter when there was little else to do. It was recognised as a worthwhile way to obtain species like the prominents, and others that did not come to sugar or were not easy to find by other methods. Pupae, being easy to pack and requiring no sustenance, were also convenient for mail-order sales. This was quite a profitable sideline then, and dealers like Reid of Pitcaple ensured that northern Scottish specialities like the Sweet Gale were among the commonest species in English collections (Tutt, 1902). Canny Aberdonian that he was, Reid even got his advertising for free, by disguising it as scientific notes about how numerous and easily obtainable the relevant species were in his area!

Today, the mercury vapour light trap, as already discussed, has largely superseded this method of fieldwork. It is no exaggeration to say that more moths would result from one successful night with the light trap than from a whole winter's diligent searching for pupae. However, not everyone has a light trap, and there are sites and situations where it is difficult to use one. Even with this new technology, certain species are still more easily found in the pupal stage than in any other. The Northern Dart is one, because of the logistical problems of trapping at night in its mountain haunts. Clearwings (Plate 98) do not come to light, and are next to impossible to breed from the egg, hence most are still obtained as pupae. Other internal feeders like Webb's Wainscot and the Bulrush Wainscot *Nonagria typhae* are usually collected in this stage too. A further advantage is that the resulting moths are sure to be in pristine condition, and there is an equal chance of them being females, which are often scarce at light.

Searching for pupae can also increase the variety of species found if the opportunity to work a new area is brief, as when on holiday. Moths not yet on the wing at that season can still be sought. One August I dug three pupae from beneath an oak tree in Perthshire. Two were Clouded Drabs, common enough in my home area. However, I could hardly believe my luck when the wing pattern of the third pupa developed and proved it to be a Merveille du Jour, a moth I had coveted for years.

Whether or not it is strictly necessary, pupa hunting can be interesting and enjoyable in its own right. The traditional technique is to dig carefully in the topsoil around the base of trees, using a garden hand fork or trowel. This

Plate 98: Large Red-belted Clearwing *Synanthedon culiciformis*. Feeding internally as caterpillars in stems or trunks, and rarely noticed in the adult stage because of their remarkably wasp-like appearance, many clearwings are most easily obtained as pupae. (Moray).

Plate 99: Pine Hawk-moth *Hyloicus pinastri* pupa. Subterranean pupae are typically red-brown and glossy, often in no more than a slight cocoon or earthen cell near the base of the food plant. (Norfolk).

relies on the fact that many caterpillars burrow immediately into the first available soil after descending the trunk, if they do not have a wandering phase first. Most native deciduous trees are productive, offering a wide range of potential species, but oak is supreme. Large isolated trees in parkland, along the edge of a copse, or bordering a glade are preferable to those in dense woodland. The quality of the soil matters too. Light, dry, friable or sandy soils are best, if only because they are easier to dig. Waterlogged soils are unpromising, and clay soils hard to work. Particularly in birch woods, the matted fibrous roots of grasses, heather and bilberry can make excavations nearly impossible. If there are tunnels signifying mole, mouse or vole activity, failure is almost certain: any pupae will have been eaten.

Care must be taken to minimise the risk of injury to the finds, though not all casualties can be avoided. Once the integument of a pupa is pierced or split, death is inevitable. First, pull up any clumps of grass growing next to the trunk, and shake the soil from their roots. With luck, a pupa will fall out too. Then lightly fork and sift the soil around the tree, paying special attention to any hollows and crevices between large, partly exposed roots and buttresses. Most pupae will be near the surface and close to the trunk, often just under the moss layer, so there is little profit in digging deeper than about 10 cm or further than 20 cm away, unless it is a particularly productive tree. Usually, excavated pupae are immediately visible, the disturbance having broken open the cell or slight cocoon in which they were housed (Plate 99). Those encased in tougher, earth-covered cocoons are liable to be overlooked, or mistaken for a lump of soil. They are sometimes more easily found by feel. Finally, replace the soil, moss and turf neatly. Afterwards, it should not be possible to tell where you have been working.

Another good place to look for pupae is behind loose dead bark. For some species, such as the Grey Dagger and May Highflyer *Hydriomena impluviata*, this is their normal pupation site on the host tree. However, a surprisingly wide range of species that usually pupate in soil or detritus can occasionally be found under dead bark. In many cases they feed on grasses or low plants, so the type of tree is immaterial, nor does it matter whether it is alive or dead. Flaking bark on fenceposts suits them just as well. The Clouded-bordered Brindle *Apamea crenata* falls into this category, as do the Bright-line Brown-eye *Lacanobia oleracea*, the Spectacle, Small Fan-foot *Herminia grisealis*, White Ermine, Common Pug *Eupithecia vulgata* and a host of others. Pulling away the bark breaks open the slight cocoon of chewed wood and sawdust to reveal the pupa. In spring, caterpillars too are often found, hiding during the day before climbing the tree to feed at night. Unfortunately, dead bark cannot be satisfactorily replaced once pulled off. It is important for numerous other invertebrates, so damage to the habitat should be kept to a minimum.

Inevitably, this type of pupa hunting is rather indiscriminate – a lucky dip. The chances of finding a particular species can be improved by concentrating efforts on its preferred tree, if it has one. Even so, most of the pupae found will be the commonest ones. Indeed, there will be days when one is grateful to find anything at all. Tree after tree produces nothing, then suddenly a little cache of several under one tree redeems the session. Even then, the rewards are rarely commensurate with the effort expended unless a value can be placed on the small thrill that accompanies each hard-won find. A glossy plump healthy pupa becomes the pearl in the oyster shell, the gleam of a gold nugget in the washing pan. Then there are other images to store in the memory: the strange toadstool or unusual beetle, the flock of Long-tailed Tits flitting through the bare branches, and the dashing male Sparrowhawk that caused them such shrill panic, the fox that nonchalantly sauntered across the glade upwind... What else is there to do in the winter anyway?

Whereas a caterpillar can be observed and released, disinterring a pupa or otherwise damaging its cocoon leaves it exposed and vulnerable. Searching for pupae therefore presupposes that any found will be kept until the moth emerges, whether that species was being deliberately sought or not. Fortunately, the care of pupae is relatively simple, so this is a good way to start rearing moths.

Chapter 9

Rearing moths in captivity

For the professional entomologist, rearing moths might be simply part of the job, studying the life history of a pest species to find ways of controlling it. For the amateur, the motivation is quite different. Rearing moths is fun. Of course, one must not say that, so it is customary to hide behind all sorts of pompous, worthy justifications. Certainly there can be an element of scientific enquiry – even for a few common species, the early stages have not been properly described and the life history is unclear. Sometimes the information gained can help with practical conservation measures for the moth concerned – though little good it did the Essex Emerald. Usually, though, moths are reared for interest and pleasure, and to provide perfect specimens for collections or photography. Unless this adversely affects wild populations of vulnerable species, there is nothing wrong with that.

Most people begin rearing moths almost accidentally after chancing upon an unusual caterpillar (Plate 100), digging up a pupa while gardening, or perhaps being offered surplus live material by a friend. If encouraged by success, they become more ambitious, until rearing is a hobby in itself. It can be very satisfying, and far cheaper than breeding horses. Caterpillars are not much different from any other livestock, or even human offspring, and the same basic principles apply. Feed them the right food, maintain a suitable temperature and humidity, give them space and air, and keep them clean. Usually they will thrive. The other stages are even less demanding.

Even so, it is best to begin with easy and straightforward species (which are not necessarily the commonest ones), avoiding those with a reputation for being difficult. The latter include most internal feeders and those that overwinter as caterpillars. Whether the relevant food plant is easily available should be another consideration. If not, a little forward planning can help. I wanted to rear the Scottish Highland race of the Dark Bordered Beauty,

Plate 100: Casual finds like this striking Poplar Hawk-moth *Laothoe populi* caterpillar, of the unusual red-spotted form, have doubtless encouraged many beginners to start rearing moths. (Banffshire).

which feeds on Aspen. No Aspen grew locally. Fortunately, the eggs over-winter, so there was plenty of time to contact a tree nursery. When the eggs hatched fully 10 months later, four sapling Aspens were producing ample tender young growth for the caterpillars. A little extreme, perhaps, but this is a most attractive moth! (Plate 130).

To observe the whole life history it is necessary to start with the egg. Eggs are hard to find in the wild, and the species that laid them is frequently unknown. However, when eggs are obtained from captive female moths, both these problems are overcome. If a female of one of the scarcer or more interesting species is caught, it is always worth considering this course of action. The following is intended only as a general guide to the rearing of most macro-moths, largely ignoring those with specialised requirements or that present unusual difficulties.

Persuading captive females to lay

Some moths lay readily in a container even without the presence of their food plant. Eggs can be found (appropriately enough) on the egg trays in the light trap, and often these are recognisable too, on the evidence of that night's catch. Generally such eggs are unwanted, but once I was delighted to find several neat strings of the Pale Eggar's, never having realised at the time that a female had been caught rather than the usual males.

More often, females need to be stimulated to lay by enclosing them in a roomy container with the appropriate food plant, or a choice of food plants if there is more than one. Not all female moths lay on leaves: some need flowers or seedheads, or dead and withered stems rather than fresh. A few Ennominae like the Spring Usher *Agriopis leucophaearia* use a long ovipositor to push their eggs into cracks and crevices in twigs, bark or dead wood, as do some noctuids. Crumpled tissue paper or muslin may serve as a substitute. It can take several days before the female co-operates, and during this time she must be fed. Fresh nectar sources like sallow catkins, Buddleia or ragwort heads can be offered, or a bit of sponge soaked in a very dilute solution of sugar or honey.

However much care is taken to simulate natural conditions, some females stubbornly refuse to lay. The Heart Moth *Dicycla oo* is said to be notoriously difficult, as is Dotted Rustic and the Ear Moth. Presumably we are unable to create in captivity the exact conditions they require. Some Geometridae are hard to persuade too, and my best efforts have failed to obtain any eggs at all from the Rivulet *Perizoma affinitata*, Marsh Pug and V-moth among others, in spite of supplying them with the correct food plant and everything they might require. Individual females of the same species do vary, some being willing and others reluctant in identical conditions. (And not just moths…)

In these situations, do not neglect the obvious. Check again that the moth is female! The sexes look very similar in some species, and I have made this error three times myself. The frenulum (Figure 7) is diagnostic, if hard to examine in a live moth. Next make certain that eggs have not been overlooked, as they can be extremely well hidden, especially if plenty of food plant has been provided. Sometimes the female's shrinking abdomen is a clue

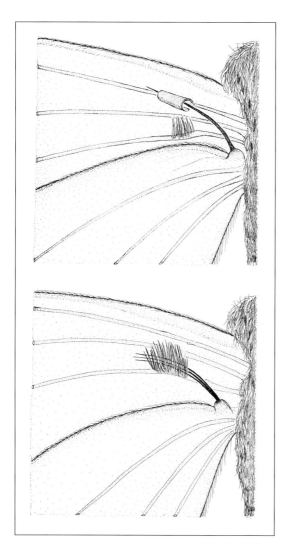

Figure 7: Sexing moths by means of the frenulum: male (above); female (below).

Plate 101: Like most moths that hibernate, the Sword-grass *Xylena exsoleta* does not mate until well into spring. Inconspicuous amongst last year's dead stems, the pair remain joined throughout the next day. (Banffshire).

that she has laid, prompting a thorough search. If there really are no eggs, one possibility is that the female is unmated. This happens occasionally, especially if the species is not numerous in the area, or she was newly emerged when found. Check that she is not constantly 'calling': if she is, it might be possible to assemble a male. Most of the hibernating noctuids like the Sword-grass do not mate until well into spring, so it is necessary to closet both sexes together until this happens (Plate 101). The wait can be as long as a fortnight, and during this time the moths feed and excrete so copiously that their cage needs regular mucking out.

If the female has apparently mated but still will not lay, transferring her to a larger container kept out of doors sometimes helps. Better still is a fine mesh net cage or sleeve, placed over growing food plant. However, there are times when nothing works. Either the decision is taken to release the female, or she dies of old age. If the latter, dissection may show that the enterprise was doomed from the start because the female was completely spent, having laid all her eggs before capture. Even fresh-looking females can be empty.

Fortunately, most attempts end in at least partial success, and some eggs are obtained. There is no need to be too greedy in any case, unless large sample sizes are needed for a special study. Twenty or thirty eggs should be ample, as rearing that many caterpillars will require a fair amount of effort. After this, the moth can be released to lay the rest of her eggs in the wild.

Care of the eggs

When first laid, eggs are commonly cream or pale green. Within a few days, they often darken or change colour, and sometimes a pattern appears. This is a good sign, indicating that they are fertile. Infertile eggs remain pale, and usually collapse and dry up within a week or two.

Eggs that will hatch soon require little care, beyond being kept at a moderate temperature and humidity, and safe from invertebrates that would eat them. If they were laid on fresh food plant, mould can be a danger when this wilts, especially in a closed container. Eggs that will overwinter present slightly more problems. They must not be allowed to dry out, but too damp an atmosphere encourages mould, so a balance has to be struck. Such eggs are best kept outside in an unheated shed. This also reduces the risk that they will hatch too early, before any food plant is available, a potentially fatal outcome.

Many eggs change colour shortly before they hatch, those of noctuids often becoming a deep purplish grey. Sometimes the embryo caterpillar can be discerned with a lens. Others, like those of the Broom-tip *Chesias rufata*, give no such warning, nor do some overwintering eggs with tough and opaque shells. It is vital to keep a close eye on all eggs around the time hatching is due, checking them at least daily. The tiny caterpillar has few energy reserves, and often dies within hours of exhaustion and dehydration if it cannot feed. Whole broods can be lost if they hatch unexpectedly and find nothing to eat.

The newly hatched caterpillar

Once the eggs have hatched, the caterpillar often eats some or all of the shell for its first meal, and should not be disturbed while doing so. The next stage is crucial: the hatchling must begin feeding on the appropriate food plant. It might be assumed that this would follow automatically, and often it does. There are occasions, however, when it seems that nothing will tempt the caterpillar to eat. This is not always a worrying sign, because some have the urge to disperse immediately after hatching, and must satisfy this need first. They climb to the top of the container, or towards the light, and will escape through the smallest chink if they can. Once past this phase, they begin feeding readily.

The food plant in the wild of almost every British macro-moth is now known, if sometimes a trifle vaguely: unspecified 'grasses' and 'low plants' appear all too frequently in the literature. Authors such as Allan (1949) also list alternatives that are acceptable in captivity. Polyphagous species should be offered variety (there may be local races or strains that prefer a particular plant), and tender young growth as well as more mature leaves. It is best not to give newly hatched caterpillars too much space, or there is a danger that they will wander far away from their food and starve. Because of their instinct to climb, make sure that the food plant reaches to the lid of the container.

If caterpillars continue to refuse food, the only solution is to experiment. Agar is sometimes used to start off difficult species. Others change their diet as they grow. I once obtained eggs of the Feathered Ranunculus *Polymixis lichenea*, and offered the hatchlings every food plant listed in the books, but nothing tempted them. In desperation, down to half a dozen survivors, I tore up handfuls of seedling weeds from the vegetable patch. That is what saved them, for amongst the weeds were a few blades of grass, and these they devoured greedily. Once half grown they ate all the plants they had refused before.

Sometimes unidentified caterpillars are found. Usually they will be on or near their appropriate food plant, or the habitat will provide useful clues. Otherwise, the majority of caterpillars likely to be found wandering will eat one or more of the following plants: dock, knot-grass (Plate 102), dandelion, clover, bedstraw, sallow, oak, bramble, and grass.

Plate 102: In captivity, Knot-grass is a universal standby for most caterpillars that feed on low-growing plants, here a fourth-instar Ashworth's Rustic *Xestia ashworthii*. Its markings are typical of noctuid caterpillars that hide in low vegetation: a shape-splitting broad pale spiracular stripe, and dark subdorsal dashes. (North Wales).

Rearing in containers

After the first hurdle is surmounted, and the caterpillars are eating well, there are several ways of supplying them with food. The simplest is to provide picked food plant in a container, and the vast majority of species can be reared without problems by this method. Fancy rearing cages are unnecessary: an ordinary jam jar works perfectly well, so long as it is not left on a sunny windowsill. It has several merits, the most important being the all-round visibility of its contents. This makes it easy to check on the health of its inhabitants, observe their behaviour, and see when the food plant needs changing. A raid on the kitchen will provide other useful rearing equipment for free. Empty plastic tubs of various sizes that once held margarine, yoghurt or ice cream are ideal. They are easily sterilised, and suffer less from condensation than glass. Line the bottom with paper towel to absorb surplus moisture, and cover the top with clingfilm, which is slightly permeable to air. The aim is to control the humidity – too low and the plant wilts quickly, too high and small caterpillars drown in the resulting condensation. The clingfilm can be held on by an elastic band, or by the original lid with the centre cut out to make a window. Very few caterpillars chew through it and escape, but the Marbled Clover will do so.

As a refinement, the cut food plant can be stood in a pot of water to delay wilting, then placed inside a larger container such as a plastic bucket. In this case, the cover should be of fine mesh for proper ventilation. A few dead twigs will act as ladders, enabling fallen caterpillars to regain access to the food plant. It is essential to guard the top of the water pot, leaving not the smallest gap, or to use porous foam blocks designed for flower arranging. Otherwise, caterpillars will crawl down the stem of the food plant until completely submerged, then drown still clinging to the stem. Remarkably, they seem to lack the instinct to turn round and climb out.

Selecting food plant

In the wild, caterpillars can and do choose which leaves to eat, rejecting any that are diseased, high in tannin, low in nutrients or otherwise unsuitable. In captivity, you must do the choosing. A good principle is to imagine you are selecting salad vegetables for your own table: pick only what you would happily eat. Avoid leaves that show signs of viruses, rust and mildew, and those that may have been exposed to pollution or sprays on road verges and at the edges of crops. Discard leaves soiled by bird-droppings. Check that no spiders or other potential predators are lurking in the chosen portion. It is surprisingly easy to introduce them to the rearing cage, likewise small caterpillars of other species. Sometimes these are a welcome bonus, but not those of the carnivorous Dun-bar, as once happened to me. Honeydew is said to cause digestive upsets and can glue tiny caterpillars to a leaf like a postage stamp on a letter.

If there is already insect damage to the plant, the level of defensive chemicals in its leaves might be raised. On the other hand, existing damage is a good sign, proving that this particular plant is palatable. They do vary: one sallow bush at my own site is almost untouched every year, while seemingly identical ones nearby have scarcely a leaf unbitten. Captive caterpillars also reject this bush. Again, one year I bred the Dwarf Pug, selecting sprigs from three spruce trees of the same commercial planting, growing side by side in the same aspect and soil. The caterpillars ate the needles of only one of the trees, declining the others even when hungry. I never did get round to marking the sprigs, so had to give them one of each every time the food was changed.

It is not safe to assume that caterpillars necessarily prefer pale young tender leaves. Often they do, but the Clouded Magpie caterpillar likes tougher, more mature ones. Larch Pug caterpillars also prefer needles that are several months old although new growth is sprouting at the same season. Much of this information is not in the books, but must be learnt at first hand: observe what your caterpillars are eating, and give them more of the same.

Husbandry

Cut food plant inevitably wilts. It must be changed before it has lost too much of its water content, or decay and fermentation set in. Frass quickly grows mould if containers are not cleaned out. Whilst conscientious attention to these tasks does not guarantee success, neglect of them invariably leads to failure. Hungry caterpillars turn to cannibalism, or at least nip each other. Contagious fungal, viral and bacterial diseases thrive in unhygienic conditions when caterpillars are stressed. A sick caterpillar always dies: there is no cure, so it is useless to call out the vet.

Ideally, caterpillars should be cleaned out and their food plant replaced just before this is palpably necessary. Better a day early than a day too late. If caterpillars are eating only a small part of the plant, such as flower buds, it may falsely seem that plenty of food is left. Note also that certain plants continue to look fresh even though they are stale. Conifers are particularly deceptive, as the needles dry out without wilting or losing their colour, and heather similarly belies its age. Fleshier plants like dock wilt all too quickly. It is easy to be caught out by a spell of hot muggy weather, which makes the caterpillars consume more, and speeds up the rate of putrefaction. Soon there is a stinking, foetid mess. The food plant may need changing daily in these conditions.

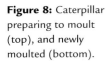
Figure 8: Caterpillar preparing to moult (top), and newly moulted (bottom).

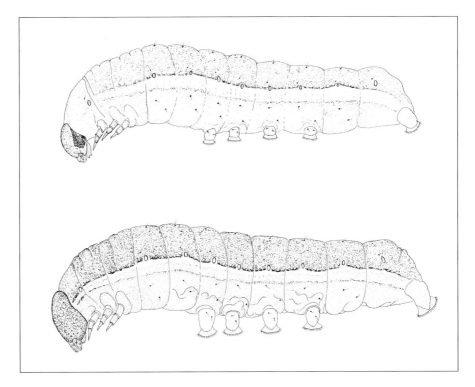

Otherwise, disturbance and handling is best kept to a minimum. Caterpillars are particularly vulnerable when very small, and should never be touched by hand. Often they respond to vibration by dropping and dangling from a silk thread, and they can be hooked up and ferried using a small paintbrush. Alternatively, fresh food can be put in with the old, which is removed as soon as the caterpillars have made the transfer. Those preparing to moult (Figure 8), often anchored on a mat of silk, must not be disturbed or they may fail to do so successfully.

Opinions differ on whether caterpillars need spraying. In the wild they would be exposed to rain and dew, but most get all the water they need from the food plant. That of the Drinker Moth is an exception, hence its name: it will not thrive unless allowed to indulge its habit. Lichens contain relatively little water, and in captivity the caterpillars of footman moths that feed on them also benefit from a good spraying, drinking greedily. I give them and their food a thorough soaking every 3 days, but make sure to provide excellent ventilation, allowing surplus water to evaporate quickly (Plate 103). Hard dry frass is often the pointer that more moisture is needed.

As caterpillars grow they need more room. Overcrowding encourages disease, which then spreads through the whole brood. It is wise to split large broods, rearing them in several separate containers. Then, if disease does strike, there is a chance it will not affect them all. If any caterpillar seems to be ailing, remove it quickly. When there are more caterpillars in a brood than you really need, do not fall into the trap of thinking that it does not

matter very much if some are lost. That is a recipe for losing them all. Ironically, starting off with only a few eggs or caterpillars is a better scenario for success because none will be considered expendable.

Rearing on growing food plants

This overcomes the problem of wilting when cut food plant is used. It is a particularly useful way of overwintering partly grown caterpillars that nibble occasionally during mild spells, as it approximates to natural conditions. Smaller plants can be grown in a pot. Primrose, Foxglove, Bird's-foot Trefoil and plantain all do well in pots, either in good light in an unheated room or a sheltered situation outdoors. Fine netting keeps the caterpillars in and predators out, assuming that no spiders, beetles, centipedes or earwigs were lurking on the plant or in the soil around its roots when it was potted up.

For larger plants including trees and shrubs, sleeving is an option. A loose sleeve of fine but tough netting (easily made from a lady's stocking) is slipped over the whole plant or a branch of a tree, encasing the caterpillars,

Plate 103:
Caterpillars reared on lichens, like the Muslin Footman *Nudaria mundana*, need special care to prevent either dehydration, or losses due to stagnant dampness. In this captive-bred 'family group', the female has laid eggs on two adjacent cocoons as well as her own. (Kincardineshire).

and the ends tied securely. Such a device is practical in private gardens, but would obviously attract attention if there is public access. Where large numbers are being reared there is considerable saving of effort. At the turn of the nineteenth century, dealers such as LW Newman used this technique to supply up to 50,000 pupae for sale each winter. On a smaller scale it can be useful for tricky species that are difficult to rear in captivity, and again for overwintering partly grown caterpillars.

Pupation

Initially, all my boyhood attempts to rear moths ended in disaster. Everything would go well at first. The caterpillar ate its leaves and grew large and plump, brimming with health. Then, just when it looked its best, it would stop feeding. Even the choicest leaves would not tempt it to eat. Instead, its bright colours faded, it shrank visibly, and soon it was lying on its back at the bottom of the jar, inert except for the occasional twitch. Sadly, I would tip it out at the bottom of the garden, wondering where I had gone wrong.

In fact, I had done everything right, except to realise that the caterpillar was ready to pupate (Figure 9). Even for experienced breeders, this is a crucial stage because weeks or months of effort can be wasted if the transition does not go smoothly. Failure to provide the appropriate conditions for a particular species can be fatal, or can lead to a deformed pupa that in turn produces a deformed moth.

Recognising a full-grown caterpillar is the first step. There is a useful rule of thumb for caterpillars of moths: the maximum length in the last instar is roughly the same as the wingspan of the adult. This works for most groups, except the burnets and the footman moths. (Curiously, it does not work for butterflies.) Once fully grown, the caterpillar stops feeding and rests while the last meal passes through its gut. Often, the final dropping is loose and wet, sometimes causing a characteristic orange-brown stain on paper lining the container. While this is happening, the caterpillar becomes shorter but not slimmer. It feels firmer to the touch if handled. In some species there is

Figure 9: Caterpillar preparing to pupate.

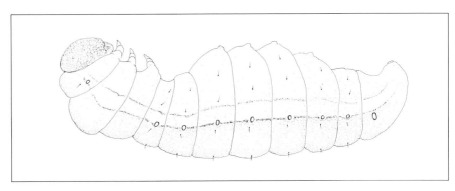

a colour change, or patterns disappear. Usually the skin appears taut and greasy. Humps and protuberances become less pronounced. Many caterpillars start to crawl urgently around the bottom of the container while these processes are taking place, and this is the moment to remove them before they soil themselves with frass.

The pupa of most macro-moths is either on or below the surface of the ground – concealed in surface detritus or leaf litter, or in the earth itself. In the wild the caterpillar carefully chooses its own site; in captivity, suitable provision must be made unless it has already started to spin a cocoon amongst the contents of its container. The pupation site of most species is given in the literature, but this information is often wrong, being based on accounts of rearing in captivity where the caterpillar had to make do with whatever was available. In these circumstances it may have had no choice but to make a cocoon amongst the remains of its food plant, the lining tissue, or even from its own frass. The conscientious breeder will provide more appropriate materials long before this point.

On suspecting from its appearance and behaviour that a caterpillar is ready to pupate, it is best to transfer it to a clean new container partly filled with a suitable medium. A small piece of food plant can be included just in case the assessment was wrong. For subterranean pupae horticultural peat or potting compost is often used as this is relatively clean and sterile. Ordinary garden soil will serve, though it may contain predatory invertebrates or pathogens unless sterilised by baking it for a while in an oven. A depth of about 10 cm of fairly firm, slightly damp peat or soil is adequate for most species. Ideally, each caterpillar should be given its own quarters, otherwise interference causing damaged cocoons and deformed pupae sometimes results. Certainly, too many should not be crowded together.

For caterpillars that make cocoons in surface debris, coarse fibrous peat again is useful, covered perhaps with a layer of dead leaves and dried sphagnum moss. Quite a few species require dead wood or bark. Virgin cork can be offered, or corrugated paper tightly rolled and bound into cylinders. Dead Elder stems with the pith partly drilled out are excellent for *Acronicta* species like the Poplar Grey. Caterpillars that spin silk cocoons, like tiger moths and tussocks, appreciate dead sprigs of heather.

Sometimes it seems as if caterpillars have rejected the arrangements made for their pupation. They crawl frantically but aimlessly around the container as if searching for something better. Generally, this is not a criticism, but merely a stage through which they (like other adolescents) must pass. Even in the wild, many species wander for several days before selecting a pupation site. During this time, considerable shrinkage takes place, enabling for example a Pale Eggar caterpillar that was 42 mm long to fit itself into a cocoon only 16 mm in length (although at first the caterpillar is bent double inside, and the final shrinkage takes place after the cocoon is completed). When physiologically ready, such caterpillars will cease their perambulations and begin to burrow or spin.

Care of pupae

Having obtained a perfectly formed pupa, it is tempting to regard success as already guaranteed. Counting one's chickens in this way is only slightly premature: nearly all healthy pupae do go on to produce moths, given a minimum standard of care. The two main dangers are mould and other fungal diseases if kept in too damp or stagnant an atmosphere, and dehydration if kept too dry.

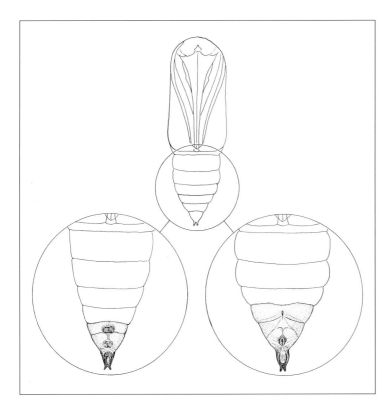

Figure 10: Sexing pupae: male (left); female (right).

Do not be too impatient to unearth buried pupae, or open cocoons to look inside. Caterpillars of many autumn noctuids and others rest for 6 weeks or more inside the cocoon before pupating, and it is vital not to disturb them during this time. Also, a newly formed pupa of any species is soft and easily injured until the chitin hardens and darkens. As a general rule, pupae formed in soil or peat can safely be excavated after a suitable interval and placed on the surface where they can be observed. They must be spaced widely enough not to touch and cause cross infection if one succumbs to disease. Check them occasionally, removing any that are obviously dead and mouldy. Pupae can also be sexed (Figure 10), either for interest or to help with further breeding plans.

Overwintering pupae are best kept in an unheated and well-ventilated outhouse, but protected from severe frost, and from anything that might try to eat them. This will also ensure that, if the moths are to be released, they will emerge at about the right time of year. Opinions differ on whether spraying with water at intervals is beneficial.

Whether pupae should be removed from their cocoons depends on the nature of the latter. If a cocoon or earthen cell is so slight that it crumbles, there is no harm in removing the pupa. But the more elaborate and strongly constructed the cocoon, the more important it is to leave it intact. Clearly, the caterpillar would not have expended so much effort unless the cocoon served a purpose. It is usually fatal to remove *Cucullia* species from their tough, roomy, silk-reinforced earthen cells, for instance. Gently shaking such cocoons can either reassure or dishearten the anxious breeder: solid clunky thuds are a healthy sign, but silence or a dry rattle suggests the caterpillar has died and gone mouldy or shrivelled up.

Emergence

Like hatching, moulting and pupation this is a critical event for both moth and breeder – the final hurdle. Success is so close, but there is still a real risk of failure if the process does not go smoothly. This risk can be minimised by providing optimum conditions.

Most pupae give a series of warning signs that the moth is soon to emerge (Figure 11). First, the eyes darken, and at about the same time the wings begin to form. Their unpigmented scales gleam through the wing cases. Then the darkest elements of the markings are laid down, and gradually the whole wing pattern appears, like a Polaroid photograph developing. The pupa often becomes uniformly dark by the end of this stage. Unless it is a species that overwinters in this state, the moth will soon emerge. First, secretions soften the chitin to help the moth break out more easily. Shortly before doing so, it takes in air through the spiracles and the abdomen lengthens, stretching the articulated segmental divisions of the pupal case. The pupa lightens as the fluid in which the moth was bathed dries out, and the wing pattern once again shows clearly through the weakened shell. Emergence is now imminent, and may be triggered by touching or merely breathing on the pupa.

These changes cannot be seen if the pupa is encased in a cocoon. In a few

Figure 11: Outward signs of the development of the imago within the pupa. From left: newly formed pupa; eyes darken and wings gleam; wing pattern begins to appear; fully pigmented; emergence imminent.

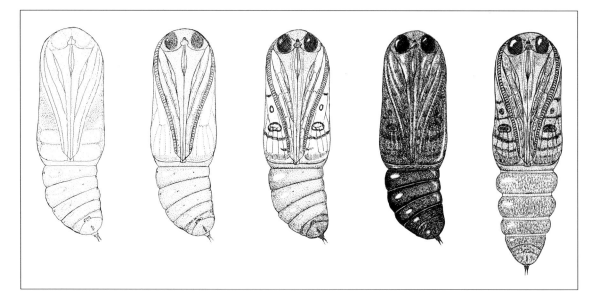

species like the Kentish Glory, the pupa itself ruptures the cocoon shortly before the moth emerges, but generally there is no advance warning. Thus it is important to transfer cocoons to an emergence cage in good time rather than risk the moth arriving unexpectedly in unsuitable surroundings. It is also hard to see the wing pattern developing in pupae that are particularly dark and opaque, although the gradual softening of the shell remains a useful indicator, and the final lengthening of the abdomen is unaffected.

Although it is possible to rear moths as far as the pupal stage using nothing more than empty food containers, acquiring a purpose-made emergence cage is well worth the effort or expense. Nothing too elaborate is required, merely a wooden box with the sides and top deliberately roughened and scored to give a secure foothold for emerging moths. Dead twigs propped inside offer moths additional opportunities to hang, as many prefer to do, while they inflate their wings. There should be ventilation openings covered with perforated zinc or an equivalent modern substitute, allowing the pupae to be laid on slightly damp peat or moss, or in the grooves of corrugated paper. A sliding front of clear Perspex allows the observer to see what is happening, and prevents the cage being so gloomy that moths assume it is night and become active, perhaps damaging themselves.

Waiting for a moth to emerge is always exciting, especially when it is a new species for the observer, and is the culmination of much effort and skill. It is hard not to feel an almost parental pride. Watching it break open the pupal case and wriggle free, then expand and dry its wings, never loses its ability to fascinate however many times it has been seen. Even the most nondescript moth is shown to advantage when it is newly minted, not a scale out of place, all its shades and colours glowing and unfaded.

Release

When a bred moth is not required as a specimen for a collection, the obvious step is to release it, perhaps after taking photographs. If the species occurs naturally in the area there is no problem. Whether the release of bred specimens, even in large numbers, has a beneficial effect on local population levels is highly doubtful, although it would be nice to think so. At least, it is unlikely to do any harm, albeit there are hypothetical dangers. Ideally, emergence will have been timed to coincide with the usual flight season by keeping stock at outside temperatures – there is little point in releasing a moth in March if it should not be out until May. Note that it is easier to speed up development if necessary with warmth than to slow it down. Between the airing cupboard and the refrigerator, there is usually somewhere in the house where the temperature is suitable. If not, the central heating can always be adjusted unless other members of the family object.

It is a different matter if the moth does not occur naturally in the area. Releasing any non-native species in Britain is illegal. Releasing a British species outside its normal range is merely considered inadvisable, since there is a slight possibility that distribution records will be thrown into confusion. In fact, it is extremely hard to establish even a temporary colony of a moth where it does not occur naturally. There seem to be few if any documented cases where this has happened as a result of casual releases. Having bred the Spinach from a stray female caught in my Banffshire garden, I duly released 13 adults into a large currant patch where its relative the Phoenix *Eulithis prunata* was resident. Although several females were known to have mated and laid eggs, no moths were seen the following year. Attempts to found a colony of the Magpie Moth in the same garden also failed completely. The site was within the British range of both, so doubtless they would have been present already had it been suitable.

Parasitoids

Sometimes it is not a moth that emerges from the pupa, but a wasp or fly. Or a caterpillar dies after the grubs that have been feeding inside it tunnel their way out. Anyone rearing caterpillars or pupae found in the wild will encounter parasitoids sooner rather than later. Occasionally even caterpillars raised entirely in captivity are affected via eggs or small grubs introduced with the food plant.

Parasitoids are so called because they invariably kill their host in the end, whereas a true parasite normally does not. Those affecting Lepidoptera are Diptera (two-winged flies) of the family Tachinidae, and several families of parasitic wasps, of which the ichneumons are the best known. Shaw and Askew (1983) give a general overview.

To the lepidopterist, it is frustrating to see a cherished caterpillar, perhaps of a scarce and desirable species, come to a horrific end, or to find a lean and sinister wasp rather than the long-awaited moth in the emergence cage. Little wonder that, in disappointment and revulsion, harsh retribution often follows. This is a pity, and very unscientific. Parasitoids are highly specialised insects, showing remarkable adaptations to their unusual lifestyle, yet they have been little studied. Opportunities to increase our knowledge of this group must be squandered by moth breeders all the time. To my shame, I myself have stamped upon a parasitoid new to Britain – a whole brood of them, in fact. Feeling guilty, when similar maggots began to devour a second pupa I sent them away to be identified. They proved to be a tachinid fly *Phryxe erythrostoma*, a host-specific parasitoid of the Pine Hawk-moth (Ford *et al.*, 2000). Shaw (1997) gives instructions for rearing parasitic Hymenoptera and stresses the need for precision when recording the data.

Chapter 10

Photography – an alternative to collecting

Photography gets ever easier. Modern cameras largely take over the technical aspects, leaving the photographer free to concentrate on aesthetics and the framing of the shot. Now, almost anyone can take acceptable pictures of moths, and the occasional superb one with a bit of practice and luck. It helps that moths themselves are very photogenic.

This means that to create a visual record of the species seen, it is no longer necessary to amass rows of dead corpses. Usually, a clear photograph will serve equally well as proof of identity for a rare or unlikely species, among macros at least. Increasingly, moth collections are composed of slides. Of course, photographs can only complement rather than supersede actual preserved specimens, which are still essential for many scientific purposes. For example, it is impossible to dissect the genitalia of a moth in a photograph, analyse its DNA, examine structural features such as the frenulum, or see the pattern and veins on a hindwing that is hidden from view. Otherwise, for the moth enthusiast, photography offers a valid and convenient alternative.

Advantages of photography

Above all, there is no need to kill the subject. This may be crucial for those with ethical objections, or when the moth belongs to a vulnerable and perhaps legally protected species. If but a single example of a scarce moth is caught during a group session, any number of people can photograph it, whereas only one could possess it as a specimen – a potential source of friction. When a moth happens to be female, it can be photographed and then kept for eggs if desired, without the dilemma faced by a collector: whether to kill it immediately, or risk it becoming hopelessly battered yet still not laying.

Plate 104: The disruptive pattern extending across both forewings of the Saxon *Hyppa rectilinea*, beautifully disguising it on birch trunks where it habitually rests, is rendered meaningless in set specimens but readily illustrated by photography. (Banffshire).

From the scientific viewpoint, the most important aspect of photography is that it records behaviour. A live moth can be shown in its natural resting posture. This is potentially useful for identification purposes, often being characteristic of a species or group. It can never be convincingly replicated in a dead specimen even if this is not set in the conventional spread-eagle position – wings and antennae curl, and the abdomen dries out and shrinks. Many other activities can be captured on film: wing-drying, mating, feeding, and warning displays, though for some of these a video camera might be best. A photograph can also place a moth in context by including background or substrate, showing far better than words will ever do how its shape, pattern and behaviour have evolved to camouflage it – or make it stand out, if an aposematic species – in its natural habitat (Plate 104). By contrast, the regimented rows of specimens in a collection, wings unnaturally stretched, stand divorced from their environment.

Photography is particularly useful for illustrating the early stages, especially caterpillars, since there is no entirely satisfactory way to preserve these without losing most of the colour, shape and posture (Plate 105). And if a caterpillar is freeze-dried, pickled or blown this destroys the chance of seeing the pupa and moth it would have become, whereas all stages from egg to imago can be recorded on film from a single individual if necessary. Truly, photographers can have their cake and eat it several times over, and still release a live moth at the end.

Finally, a collection of photographs is easier to store. It takes a fraction of the space required for a cabinet of the same number of set specimens. Although slides and photographs obviously require careful handling and storage to protect them from dust, damp and scratches, they are less liable to damage and deterioration. There is no need to worry that they will be attacked by mites and museum beetles, which can devastate the contents of a cabinet if they gain entry and repellents such as paradichlorbenzine have not been kept up to strength. Set specimens are also vulnerable to slight knocks and jarring

Plate 105:
Caterpillars, and their association with their food plant, can never be preserved as specimens in a lifelike manner. The remarkable camouflage of the Wormwood Shark *Cucullia absinthii* caterpillar on Mugwort, accentuated by its crooked posture, is best captured on film. (Norfolk).

that snap off brittle antennae and abdomens, while fluctuations in temperature and humidity cause wings to spring and droop. Stored body fat migrates, turning specimens greasy and combining with brass pins to form verdigris. Even with the best care, colours fade sadly or change with time, and the bloom of the freshly emerged moth gradually disappears. Today, the dyes used in high-grade colour film are remarkably stable, especially if kept away from sunlight, and can outlast moths' natural pigments (Plate 106). In 1994 I bred the Sussex Emerald and several were added to a collection since they could not be released at an appropriate site. Five years later the set specimens had lost all trace of their original colour, a most beautiful shade of leek-green, but the transparencies were unaltered.

Plate 106: The pigments of green moths like this Large Emerald *Geometra papilionaria* fade within days in the wild, and within a year or so in a collection, whereas modern colour films are increasingly stable. (Banffshire).

Equipment

Moths are very small compared with everyday subjects like landscapes, buildings and people, so photographing them requires slightly more specialised equipment. Somewhat illogically, the technique is termed macrophotography (microphotography being used when the subject is even smaller).

A professional, familiar with sophisticated equipment and techniques, will no doubt consider the following account to be laughably basic. It is intended for the beginner. Not being an expert on cameras, I can only describe the equipment that has proved reliable and satisfactory for me. All the photographs in this book were taken with a Pentax ME Super SRL (single lens reflex) camera (now no longer made) fitted with a 1:4 100 mm Macro lens, with extension tubes added when required. It is not even mine, but belongs to my wife. Tired of repeated requests to photograph moths, one day she thrust it into my hands with forceful instructions to do it myself.

Any good 35 mm SRL camera should be suitable for photographing moths, providing that it is only semi-automatic. Avoid fully automatic cameras that insist on setting the film speed, selecting the aperture, focusing, and deciding when to employ flash. They will not give the best results. Besides, it is infuriating to be overruled by a machine. Choose instead a camera that allows you to pick any settings you want, then controls the shutter to give the appropriate exposure time.

When using natural light, this exposure time will almost always be too long for sharp hand-held shots (several seconds is not unusual), so a good tripod is essential to prevent camera shake. It is worth investing in the best you can afford. Go for strength and rigidity rather than lightness: the heavier a tripod, the more solid it will be. Weight is of course unwelcome if the equipment is to be carried for long distances in the field. In this case a lighter version may be preferred, adapted so that a handy large stone can be suspended in a net or bag centrally between the legs to give it extra stability. The more expensive versions offer a choice of camera heads. A ball and socket joint is the most versatile, allowing rapid switches between horizontal and vertical formats, or any angle in between. Some models allow the telescopic centre column to be reversed, so that the camera can be positioned almost to ground level if required, very useful because moths often rest low down.

Film

Decide whether you want prints, or transparencies that can be mounted as slides. The former require colour negative film, which when exposed reverses the true colours of the subject so that blue appears as yellow, green as magenta, and so on. This negative forms the intermediate stage in the production of a colour print. Colour positive film, on the other hand, records the tones of the original subject directly.

Because there is no intermediate stage, and no overall colour mask as with negatives, transparencies undoubtedly give the best results. All the shades and nuances of the subject are faithfully reproduced – indeed, at times colour positive film can be almost too accurate, being more sensitive than the human eye to variations in the quality of natural light at different times of the day. Publishers prefer transparencies for colour illustrations, and of course they can be projected as slides during talks or lectures, or for personal enjoyment. Finally, reasonably good prints can be ordered from favourite slides relatively cheaply, giving the best of both worlds.

Having invested a substantial sum in camera, lenses and tripod, then expended considerable effort to find or breed the moth to be photographed, it is false economy to skimp on film. Use a recognised and reputable brand, not one barely adequate for cheapo holiday snaps, then send it to be developed at the maker's laboratory, not to the local chemist. A fine grain film will give the necessary sharpness and detail, but requires a longer exposure time. That is rarely a problem if a tripod is being used anyway. The speed rating of films is given as an ASA (American Standards Association) or ISO (International Standards Organisation) number; both use the same scale. The lower the number, the finer the grain and the slower the film. It may be best to begin with ISO 100 film, graduating to ISO 50 with experience, and perhaps to a more expensive, professional grade. The resulting improvement in quality is tiny, but perfectionists will settle for nothing less. Bulk purchase from the manufacturers affords big savings, and unused rolls should be stored in a refrigerator to retard deterioration. All the photographs in this book were taken on Fujichrome, now the overwhelming choice of nature photographers. Everyone sees colours differently, but to my eye Fujichrome Velvia (ISO 50) best captures the infinitely subtle, delicate shades of moths.

Camera settings

Obtaining a sufficient depth of field is one of the main technical problems to overcome when photographing moths. Depth of field determines which part of the subject is in sharp focus. For close-up work, this zone may be only a couple of millimetres wide, depending on the aperture used (the size of the hole in the adjustable diaphragm through which light passes to reach the film). The smaller the aperture, the greater the depth of field – but the longer the exposure time because less light is being allowed through. Apertures are adjusted by a series of stops, and measured by the 'f number' system whereby the larger the number the smaller the opening.

Use the smallest aperture you can. For moths, I do not like to drop below f 16, which gives adequate depth of field for almost all shapes and sizes. Where conditions allow, f 22 or f 32 are even better. With natural light and fine grain film, this frequently means that the exposure time will be several seconds, even up to half a minute. If the camera is on a tripod and the subject is not moving, this does not matter at all, although it may be unnerving at first. The resulting image can be exceptionally sharp with exposures of this length because the momentary slight vibration caused by the operation of the shutter has a negligible effect. However, there are times when depth of field must be sacrificed to the need for a quicker exposure, for instance if an intermittent breeze is rustling the foliage where a moth is sitting, or a lively caterpillar will not sit still for long, like a fidgety small child.

Even with a small aperture, the depth of field may be too shallow to get the whole of a

Plate 107: Moths like the Lunar Thorn *Selenia lunularia* cause depth of field problems for the photographer. Here, the moth is back lit to show the transparent central lunules in each wing, perhaps suggesting cracks in a dead leaf. (Banffshire).

moth in sharp focus. Carpets, waves and other moths that sit relatively flat should not be a problem if photographed square-on, but more bulky species like noctuids need greater care. A frequent mistake is to get the wings in focus, but not the thoracic fur and crests, which appear blurred and fuzzy, spoiling the picture. As a general rule, the nearest parts of the moth must be sharp, but it does not matter so much if a distant wingtip gradually drifts out of focus – sometimes this actually provides a bit of perspective, emphasising the structure. Hardest of all to photograph in close-up are moths like the Lunar Thorn (Plate 107), which have such a three-dimensional resting posture that the depth of field cannot possibly cope. In these cases, make sure that the thorax, head, and especially the eye of the moth if it is visible, are sharp enough to hold a viewer's attention so that slightly out of focus wings go unnoticed. However, while the depth of field problem can usually be overcome, it does impose restrictions on the photographer. All too often the subject, whether moth or caterpillar, can only be photographed successfully from one particular camera angle, leading to an unwelcome sameness in all the shots.

Exposure

A semi-automatic camera will itself calculate the correct exposure time, using its own built-in light-sensitive cell to measure the amount of light actually passing through the lens (TTL metering). Normally, this produces an excellent result, but there are certain conditions where it will not. The camera simply measures the amount of light reaching the film, without regard to the intrinsic tone of the subject itself. A sheet of medium grey paper photographed in close-up will be correctly exposed, and come out as grey. But a sheet of white paper or black paper will also come out as grey, because the camera will under-expose and over-expose them respectively, as if the first were simply a brightly lit scene and the second a dim one. However, a photograph of black and white chequers will be correctly exposed as they give off a medium amount of illumination.

It follows that certain subjects will mislead the camera. The background chiefly influences the exposure time, because it makes up the greater part of the picture. Most resting moths are roughly triangular and, by the laws of geometry, a triangle contained within a rectangle can never be more than half the latter's area. Allowing for a reasonable margin all round, and projecting legs and antennae, even a moth that seems to fill the frame will comprise only about a third of the photograph, and often much less. Particularly light or dark backgrounds will therefore cause the photographer problems unless measures are taken to overrule the camera's exposure meter. This is known as compensation or correction.

Experience will teach when compensation is necessary. A moth on a very light background – chalk or limestone, Silver Birch bark, a whitewashed wall, or even a clean egg tray – tends to be under-exposed (too dark) unless the camera is overridden. A moth, especially a white one, on a dark background will be over-exposed (too bleached and pale) without compensation. Note that the mature foliage of trees and many other plants such as bracken is a surprisingly dark shade of green, especially on the upper surface, and does not make a good photographic background. Wet rocks, bark and fences are also dark.

The camera will have a dial that allows the automatic exposure to be corrected for unusual lighting conditions. However, my own preference is to achieve this by altering the speed rating setting instead (fortunately my camera allows me to do this), as it gives much finer gradations. When using a roll of ISO 50 film, I shoot it at anywhere between ISO 36 and ISO 80, over-exposing and under-exposing it respectively as desired. Indeed, I normally set the dial at ISO 64 to obtain full rich deep colour, but this may be a personal preference. A transparency that is fractionally too dark can be corrected if used for colour illustrations, whereas one that is too washed out cannot. Pale moths that lack a strong pattern, like emeralds and waves, routinely benefit from a degree of under-exposure, but very dark moths require the opposite treatment else detail is lost. Be sure to return the settings to normal afterwards – it is all too easy to forget!

Lighting

Daylight is best for photographing moths. However, full direct sunlight should be avoided, except for diurnal species like the burnets. Nocturnal moths avoid the sun, resting in shade, and their muted colours and intricate patterns have evolved accordingly. They are not meant to be seen in a harsh bright glare. Nor will moths happily sit and pose for long in strong sunlight, quickly crawling or flying away to find a shadier resting place. Unfortunately, photographs taken in shadow have a cold bluish cast from the blue of the sky, so this is not a suitable solution unless a correcting filter is used. Though the strength of the sun diminishes towards evening, the light then has an orange cast, which registers much more strongly on film than in the human eye. White moths are particularly affected. It follows that clear sunny days are not good for moth photography. Instead, bright overcast or cloudy conditions give by far the best results, i.e. the truest colours, and fewest problems with exposure, contrast and balance because of the absence of harsh shadows. Moths are more co-operative too, especially if it is cool.

An alternative is to use flash to supplement natural daylight. Many photographers do, and achieve excellent results. Even with the smallest aperture,

Plate 108: The Light Emerald *Campaea margaritata*, as photographed in direct light (top), looks far too pale for effective camouflage amongst foliage. However, since it normally rests on the underside of a leaf it would be viewed against the light by a searching bird, making it much harder to spot (below). (Sussex).

the shutter speed is quick enough for hand-held shots, doing away with the need for a tripod. However, to me, there is often something not quite right about a photograph taken with flash. Either there are harsh shadows, or the subject is too uniformly lit, appearing flat. Even if the colours of the moth are reproduced correctly, and this is not always the case, the texture seems wrong. Artificial lighting has produced an artificial effect. I prefer to use flash only when there is no alternative, as when attempting to photograph moths at night.

A further advantage of natural light is that it enables the photographer to see precisely what the result will be (Plate 108). The colours and contrast of the markings on a moth's wings often change depending on the angle of the light, as with shot silk, and it is impossible to predict how they will register if flash is used. The sheen of the wings may be characteristic of a species, and with natural light the camera can be angled especially to capture it (e.g. Plate 46). Then there is the technique of modelling: using side lighting to emphasise structural form by creating shadows and relief. This works particularly well with pale, lightly marked moths that would otherwise appear rather featureless (e.g. Plate 33). Although flash can be offset to the side, and diffused or bounced to give a similar effect, the exact result is unpredictable, and will only be seen after the film is developed.

Photography in the field

Day-flying moths can be photographed in much the same way as butterflies, although this is slightly more difficult since they are smaller, and generally more active (Plate 109). Few moths bask in the sun with wings attractively spread; many do not feed, others quiver while feeding, like the

Silver Y. However, the commoner burnets make easy subjects for beginners, especially when several are grouped on one flower head. Usually there is no time to set up a tripod, so very slow shutter speeds are out of the question, and it may be better to use a faster film. A monopod (or a tripod used with unopened legs) will help to steady the camera. Even so, expect a high percentage of wasted shots with this type of photography.

Resting moths are a different matter. They are unlikely to move unless disturbed (although some carpets are very skittish), so there is ample time to erect a tripod, meticulously check the focus and settings, and wait if necessary until the light is at its best. On a field outing to the Culbin Forest in Morayshire, I found a perfectly fresh Kentish Glory resting on a pine sapling. Just at that moment a squally shower blew in from the sea. After setting up the equipment, I crouched for 20 minutes with a plastic carrier bag over the camera to protect it from the rain, snatching a shot whenever conditions momentarily abated. Finally satisfied, I caught up with the rest of the group, rejoicing at having photographed such a wonderful moth, only to find that I had missed something even more spectacular – a topless female jogger. Life can be so unfair!

Moths found on rocks, fence posts and tree trunks make convenient subjects for close-up work with a tripod (Plate 110). Where a light trap has been run the previous night, an early morning search of nearby walls and palings is often productive. Especially in a garden, there is scope for thought, care

Plate 109: Day-flying moths that do not feed, like the Wood Tiger *Parasemia plantaginis*, present fewer opportunities for the photographer. Note also how its pattern is disruptive as well as aposematic, while the yellow hindwings (flash coloration) make it harder to follow in flight. (Banffshire).

Plate 110: When photographing resting moths, time is rarely a problem. On uneven boggy ground, rock pillars were built to support the tripod during the 8 seconds exposure needed in the aqueous Highland light for this mating pair of Yellow Horned *Achyla flavicornis scotica*. (East Inverness-shire).

and thorough preparation. If necessary, a small mirror or sheet of aluminium cooking foil can be arranged to throw more light on the moth. Obtrusive vegetation can be bent or tied back, and different camera angles and compositions evaluated. Thus the margin of error is reduced, with less wastage of expensive film.

However, in the field the photographer is always at the mercy of the elements, often trying to produce good results under far from ideal conditions. It is hard to photograph moths resting on vegetation if the slightest breeze is rustling the leaves. Patience is needed to wait, listening for a lull. Refusing to admit defeat, I once used pegs and guys to tether a swaying young poplar, so as to be able to photograph a mating pair of Puss Moths on a slender branch. Innocent passers-by may agree to act as a windbreak, thinking it safest to humour anyone making such an odd request. But on very windy days photography can be impossible, as the trunks even of large trees flex right down to ground level, imperceptible except through a camera lens. Rain and bad light frequently stop play. Nor is the subject always co-operative: having searched long and hard for a good moth, it is frustrating to see it fly away. How tempting, then, to try and improve the odds by photographing moths under the more controlled conditions of a studio.

Studio photography

Any room that receives good natural light will serve as a studio: a conservatory is ideal. Preferably, the walls and ceiling should be painted matte white, to avoid colour casts. Very little specialised equipment is needed, although a lamp fitted with a daylight-type bulb is useful when natural light is poor. Diffusers and reflectors to modify the angle and tone of the illumination are added refinements that can be made out of simple materials like stiff white cardboard.

With such advantages, and a little practice, perfect results should be obtained almost every time. At least, every photograph should be technically perfect: in focus, suitably lit, correctly exposed, free from camera shake, and sharp down to the very grain of the emulsion on the film. Whether it is also aesthetically pleasing and scientifically accurate is a different matter, and far from guaranteed.

Then there are the ethical considerations. Long before taking up the hobby, I was flattered to be invited by a well-known nature photographer to accompany him on a field trip. Having drawn his attention to a handsome large Emperor caterpillar by the side of the path, I was surprised when he promptly picked it up and transferred it to a nicer clump of heather nearby. Was that not cheating? He was quite unrepentant, explaining that it would not have made a good photograph in its original position. Now, he was confident that it would.

At the time I was unconvinced by his argument that the job of a photographer was to obtain the best possible picture of the subject, employing whatever techniques and equipment would help to achieve this. Gradually I have come round to his point of view. All photographs are to some extent artificial. The very act of selecting a subject and framing it within the confines of the picture is the first departure from reality. Then there is the depth of field of the lens, which for close-ups differs from that of the human eye, creating a pin-sharp image of the subject against an increasingly out-of-focus background. And humans have binocular vision, but the camera does not. Next, light intensity is manipulated by the use of flash or a long

Plate 111: This Gold Spangle *Autographa bractea* was photographed in the wild exactly as found – except that (as in nest photography) obstructive vegetation was carefully tidied away. (Banffshire).

exposure, so that for instance a moth that was hiding in a dark crevice on the shady side of a trunk appears in the photograph to be brightly lit. Already, the simple mechanics of photography have begun to create an illusion, without any deliberate manipulation. Even with genuine, taken in the wild shots, some tinkering is often employed: perhaps vegetation obstructing the lens (or concealing the moth?) is bent aside, a wing realigned with a puff of breath, and any discordant elements that might spoil the picture are tidied away (Plate 111).

When does the legitimate employment of such technical expertise – all the little tricks and dodges of the trade – cross the boundary into subterfuge and deceit? How far is the photographer justified in faking it, giving the impression that a carefully set up studio shot was taken in the field? Perhaps the only real sin is to be scientifically misleading. It would be wrong, for instance, to illustrate sites chosen for egg laying, pupation or hibernation with photographs of captive stock that had to make do with much more limited opportunities, without making this clear. Tremewan (1985) rightly stresses that his photographs of burnet moth cocoons were taken in the wild: they would be far less meaningful otherwise. And of course it is unfair to enter studio portraits for a competition if the rules state otherwise, though doubtless it happens. The photographs in this book are a mixture of field and studio shots, or a grey area in between; I hope it is impossible to tell which.

Studio photography is the obvious solution if the moth in question was caught in a light trap or bred, or a caterpillar was reared from the egg. If all that is wanted is a voucher photograph to confirm identity, there is no need to be too elaborate. The moth can be photographed sitting on the egg tray or in a pill box, the caterpillar in the breeding cage. While this approach is honest, the result is rarely aesthetic. The photographer has failed to do his subject justice. With a bit more effort, a voucher photograph can make an attractive picture too. And there is a further element to consider: many species will only adopt their natural resting posture if placed on the appropriate substrate. A moth that rests with wings tightly clasped around a stem (Plate 112) cannot do this on a flat surface, nor will the reasons for its shape, colour and markings be so obvious. Others seek out a crack or hollow in bark, or the underside of a leaf. Caterpillars position themselves carefully, aligned perhaps at the correct angle to mimic the stalk or central midrib of a

Plate 112: A bred Swallow Prominent *Pheosia tremula* has arranged itself on the twig provided, enabling the studio shot to capture the natural resting posture of a moth rarely seen in the wild except at artificial light. (Banffshire).

leaf. Often, a moth or caterpillar will not settle down and pose nicely if it is unhappy with its surroundings. By providing a background that looks, smells and feels right, maybe the photographer is trying to deceive the subject rather than the viewer.

This presupposes that the habits of the moth are known. I find it very difficult to photograph a bred species I have never seen in the wild, and aim for nothing more than a simple portrait on a relatively safe, neutral background. It is much easier to re-create in the studio an interaction between moth and habitat that has been witnessed many times in the field. Often such a set up requires considerable forethought, effort and attention to detail, so it can hardly be considered a lazy option or a short cut. Once I carried a large lump of granite back from a mountain summit, in case the lichens there differed from the ones on rocks lower down. A piece of Scots Pine bark was earmarked for a moth that was not due to emerge for many months. Always be on the lookout for useful props, especially as they should have a short life – discard them before they become too recognisable through overuse. Other moths rest on foliage, often fairly indiscriminately. It is safe, if unimaginative, to pose them on their food plant. They are likely to be more compliant as well. Otherwise, ubiquitous plants like bramble and dock make useful alternatives.

However, certain backgrounds should be avoided, as the result is rarely pleasing, either in the studio or in the wild. Most building materials fall into this category including sawn or planed wood, unless old and weathered; anything painted, varnished or whitewashed; brick, concrete, breeze block and pebble-dash, unless mossy and lichenous. Ornamental plants and garden flowers tend to look out of place, except perhaps for garden moths. On the other hand, too contrived a background, such as a piece of bark patterned exactly like the moth, is sure to defeat its object by arousing suspicion. The setting should also complement the moth, not overwhelm it: a rather dull moth needs a background that does not upstage it. Finally, avoid exposure problems by not overdoing the contrast in tone between subject and background, and avoiding harsh shadows.

Having created the setting, the next step is to persuade the subject to cooperate. In the past, moths were often chilled or anaesthetised beforehand, or worse: sometimes they were very obviously dead. Even some well-known photographers stooped to this chicanery, yet their work was used to illustrate reputable guides. This greatly encouraged my own early efforts – surely anyone could do better than that? By all means store moths in a refrigerator overnight to keep them quiet, but then allow them to warm up and their eyes to adjust to daylight if anything more than a voucher photograph is required. A groggy moth never looks right. It is worth the risk of losing one occasionally because it is too lively in exchange for the chance of better pictures.

Persistence and a steady hand are needed to coax a moth onto the chosen setting, preferably without waking it up too much. Species that rest openly will usually co-operate, especially if the weather is cool. Females are generally easier to work with, being more sluggish than males. Moths that normally rest concealed, like the darts, often shun the light. For these, the best that can be hoped for is a basic portrait, but species like the Dotted Rustic rarely oblige for long. Some geometrids can be very difficult, fluttering to the nearest window (and into any cobweb that the maid has missed), but they will often settle on foliage held against the glass. Not infrequently, one is lost from sight. I once tried to pose a Scorched Carpet on some Traveller's-joy, a frequent resting place in the wild, but clumsily disturbed it. The moth flew wildly round the room, then disappeared. The most thorough search of walls, ceiling, curtains and furniture drew a blank, yet the moth could not have escaped. An hour later, I was on the point of giving up, then thought of the one place that had not been checked. Sure enough, there was my Scorched Carpet, displaying itself beautifully on the Traveller's-joy.

However striking or plain the subject, exactly the same degree of care is required. Inevitably, some moths photograph better than others. Those that impress in the flesh because of their sheer size and bulk, like the Goat Moth and some of the hawk-moths, can look relatively unspectacular on film, where scale is less apparent. Gaudy species like the Garden Tiger may disappoint too: a photograph adds nothing to what the naked eye has already seen. Conversely, the smaller and duller cryptic species often turn out surprisingly well. The transparency, especially when projected as a slide, reveals delicate tones and complex patterns that otherwise escape notice. Even pugs can look attractive.

Caterpillars bring their own problems. Those that feed openly will be happy to pose on their food plant, but those that remain hidden during the day may not tolerate being exposed to view (Plate 113). Not infrequently, I have been forced to give up trying to photograph them. Shorter exposure times are necessary for caterpillars that are constantly crawling, twitching or eating. Even an apparently resting caterpillar is never absolutely still because of its internal muscular activity: the heart is

Plate 113:
Caterpillars that shun the light are difficult to photograph. This Crinan Ear *Amphipoea crinanensis*, exposed in its messy boring in Yellow Flag, was too disgruntled to pose for long. (Moray).

pulsing, food is travelling through the gut, air is being pumped in and out of the spiracles. Tiny and imperceptible though these movements are to the naked eye, with a long exposure they can register on film, making the image fractionally blurred.

Moths must be photographed immediately, in case they damage themselves. With caterpillars, the temptation is to put it off until another day – there is not enough time, the light is unsuitable, or film is in short supply. Anyway, the caterpillars will grow bigger if left a few days. It is very easy to be caught out, and find that they are already preparing to pupate. Several times I have reared broods from the egg solely to photograph the caterpillar stage, then never actually got round to it.

Taking better pictures

Mere technical competence, and the right equipment, will automatically result in some very acceptable photographs. At first, these will give much pleasure, but it is human nature to strive for improvement. One way of achieving this is to develop the ability to 'see' a shot – the best angle, the most harmonious composition, the right magnification, the correct lighting – and, of course, an attractive subject. However many times you have photographed a particular species, never pass over a chance to do better, to achieve the ultimate shot that epitomises it. Some of my own favourite pictures were taken almost as an afterthought: a Yellow Shell disturbed while gardening, yet another view of a Mottled Beauty, found when searching for much scarcer species. Try unusual angles: standard full-frontal portraits of moths are now two-a-penny (Plate 114). Aim to illustrate interesting behaviour: moths or caterpillars actually doing something (Plate 115). Experiment rather than take endless routine shots. Consider photographing moths at night.

Conversely, learn to recognise what will never make a good picture, however much care and skill is employed, and do not waste expensive film. A worn or damaged moth looks much worse in a close-up, although a torn wing can often be concealed in a profile shot. Not that anyone would

Plate 114: Inevitably, standard portraits of moths tend towards sameness. Look for something different if the opportunity arises, like this snatched shot of a Narrow-bordered Bee Hawk-moth *Hemaris tityus*. (Moray).

refuse to photograph a once-in-a-lifetime species like a Clifden Nonpareil *Catocala fraxini* just because it had a few scales missing: at least it would provide a space-filler, the equivalent of a philatelist's creased, three-margin Penny Black plate XI, or repaired £1 Seahorses. (It would be interesting to know how many moth enthusiasts are stamp collectors too.) While a couple of voucher photographs are always worth taking, nothing is more depressing than frame after frame of the same unattractive moth, none of them any good.

One way to make photographs more useful as identification aids is to use the same magnification for moths of similar size, so that the images are directly comparable. For instance, all waves, pugs and medium-sized noctuids can be photographed at exactly the same scale by setting the magnification, then moving the camera until the subject comes into sharp focus. The actual size of the moth in the picture can be measured using a photograph of a section of white plastic ruler, taken at the same magnification.

Labelling

A collection of transparencies or prints without data is merely ornamental. As with a collection of specimens, much of the value is in the labelling. At a minimum, this should include species, date and locality, preferably with a map reference giving at least the 10 km square. A note of the circumstances of finding or capture will add interest. For bred specimens, the origin of the parents should be given.

Those with good memories might think this unnecessary. After all, how is it possible to forget when and where a scarce, long-awaited species was caught? Time plays tricks, however, and one day the precise details may be important. In Sussex, I was shown an album of photographs of numerous fine moths. Its owner spoke of his hopes next spring of finding the Pale Pinion for the first time, and adding it to his portfolio. This was strange, since the album already contained an excellent picture of that moth. He had mistaken it for a much commoner species. Not thinking the data of any value, he had not bothered to write the information on the print, so could not be entirely sure when and where the moth had been caught.

Finally, an efficient storage system is required, enabling photographs to be easily traced. Filing them in the classified species order is the most logical approach. There is no point in having superb photographs of a particular species if they cannot be located.

Chapter 11

Making a scientific contribution

This is not compulsory. It is quite acceptable simply to enjoy looking at moths. For many, however, the chance to add to our store of scientific knowledge brings interest and purpose to their hobby. Amateurs have always been prominent in the study of moths, perhaps because until recently it offered few career opportunities. As yet, there is none of the dichotomy between amateurs and professionals that is increasingly affecting ornithology. Plenty of scope remains for avocational workers (as they are sensitively termed in North America to avoid any disparaging connotations) to provide real and valuable contributions. Anyone capable of making accurate observations and keeping clear records can do this.

Site lists

Compiling a list of the species found in an area is the most basic form of recording. Obviously, such lists are most valuable in under-worked parts of the country, where the moth fauna might be almost unknown. An unusual reason for the choice of a holiday destination, perhaps, but it can be rewarding. However, up-to-date lists are needed in well-worked areas too. Distributions change, ranges ebb and flow, new species colonise and spread, others die out or become more localised. Populations fluctuate, common moths becoming scarce, and formerly scarce ones common. Old lists are soon obsolete as guides to current status, but become valuable histories instead.

First, the site must be clearly defined: vagueness can make records unusable if they are to be plotted on a map. A grid reference is essential, giving at least the 10 km square. Many counties now have a tetrad atlas, so here a four-figure reference (that is, two code letters followed by four numbers) will be necessary. Ordnance Survey maps give instructions on how to obtain a grid reference. If unsure, ask another local recorder to check your calculations. Even experienced observers sometimes make mistakes: I have been sent locations for the Mountain Burnet that are many miles offshore in the Moray Firth. The relevant vice-county should be given too, as this cannot always be determined from a map reference. Banffshire, for instance, extends over at least 32 different 10 km squares, but only four of these are entirely within the vice-county boundary.

It is best to list species in their order of classification rather than alphabetically. This will show at a glance whether the site is particularly good or poor for certain groups of moths, and which expected species have not yet been noted. Also, if you carelessly write down White-spotted Pinion instead of Pinion-spotted White, and neglect to include the scientific name, the error will be clear from its position on the list.

Numbers and status

As mentioned already, simple lists can be misleading. They offer no way of telling whether a species is abundant, or has only been recorded once, as a stray, in 50 years. More information is required. An annotated list, giving an indication of the numbers seen, is far more meaningful.

A generalised assessment of status is better than nothing. Everyone understands roughly what is meant by terms such as common, occasional and rare. Unfortunately, this approach is vague and subjective. Some observers use 'common' to mean 'widespread', others to reflect numbers. One worker might describe a species as 'fairly common' if two or three are seen nearly every year. Another might apply the same description if half a dozen could be expected on a good night. There is no substitute for hard data: actual counts of numbers seen, by place and date.

Such counts will of course be influenced by the methods used, plus the time and effort expended. Somehow, they need to be put in context and standardised so that direct comparisons are possible, either with other sites or between different years. Light trapping offers one opportunity, where the same design of trap is used and a bulb of similar strength. It is particularly reliable for year-to-year comparisons at the same site, since potential variations in effectiveness because of different locations and habitats are eliminated. The weather also has an influence, and in a good summer there will be many more suitable nights than in a poor one. Thus adding up the nightly catches to give an annual total for a species is not necessarily reliable. My own preference is to use its highest total in a single night as the index for the year, since bad weather is unlikely to last for the whole of the flight period. This method also eliminates problems caused by the unknown percentage of retraps in consecutive catches.

As hundreds of moths may be trapped on the best nights, counting them accurately can be a problem. I find that speaking into a 'pocket memo' dictating machine is the answer. It enables the night's catch to be recorded quickly, without fumbling with pen and paper while moths are escaping, then transcribed at leisure. Conventionally, the date when the trap was put out is used, not that of the morning after. Thus moths caught on the night of sixth/seventh of the month are put down for the sixth, even if they are known to have arrived just before dawn. Brief details of the weather – temperature range, wind speed and direction, cloud cover, whether there is a moon – will add interest, especially if migrants are caught.

Sugaring also lends itself to systematic recording if an identical routine is followed every time. At my own site, the same 24 fence posts and an unchanged recipe have been used consistently for 10 years now, providing trustworthy data. Again, the highest total in a single night seems to be the most reliable measure of a species' fortunes in a particular year. This will slightly underestimate the true figure, since it does not allow for any turnover of individuals, but such biases do not matter if they are consistent. In the dark, a memo machine is again handier than a notebook, especially when the weather is damp.

Counts can also be standardised by distance covered or time spent searching: 17 seen in 1.5 km, 23 disturbed in 50 minutes, and so on. What matters is that the counts are reproducible. When the old authors say that a species 'abounds' (as was their habit) we cannot really tell whether they were used to seeing tens, hundreds or thousands at a time. We must ensure that our own records do not cause future generations similar problems.

Handling the data

While researching for this book, I contacted numerous observers with requests for information. Most, as expected, were extremely co-operative, delighted that their records were being put to use. Sadly, from both our points of view, a few were unable to oblige. Yes, they had the data I needed – it was all in their notebooks – but extracting it would be a formidable task. Perhaps they would get round to it one day, when they retired ….

Clearly, some system of sorting and storing records is essential, before the mass of information becomes unmanageable. Some observers create an index for their field notebooks, listing every mention of a species, and each visit to a site. I prefer to summarise my records for the year, giving first and last sightings, and the maximum total on any one date, which usually tallies with the peak of the flight season. No doubt others will prefer their own methods, perhaps taking advantage of the various computer programs now designed especially for biological recording. However, a sample from my own 1996 list is provided below.

Coxcomb Prominent *P. capucina* NJ5755: 8.vi – 27.vii; max. at mv 3 on 12.vii; larva on rowan.

Pale Prominent *P. palpina* NJ5755: 31.v – 3.vii; single males at mv on six dates, two on 24.vi; female egg-laying at dusk on grey sallow, 24.vi.

Vapourer *O. antiqua* NJ5755: female on cocoon on birch, 20.x., but no eggs laid. Only 2nd record for site.

Dark Tussock *D. fascelina* NJ5755: very small larvae (8 mm) on ling from 24.iii; full-grown larva on 21.vi; two males at mv on 8.vii then female on 12.vii. NJ2836: large larva at 320 m asl, another at 700 m asl, NJ2635 on 2.vi.

Garden Tiger *A. caja* NJ5755: 8.vii – 21.viii; max. at mv 17 on 21.vii. Average year.

White Ermine *S. lubricipeda* NJ5755: 13.vi – 19.vii; max. at mv only 5 on 12.vii – a poor year.

Ruby Tiger *P. fuliginosa* NJ5755: larvae sunning from 5.iii, but no moths seen. Larvae also in NJ2836 and NJ4668.

Heart and Dart *A. exclamationis* NJ5755: 10.vii – 9.viii; singles on six dates, two on 29.vii – a good year.

Flame Shoulder *O. plecta* NJ5755: 4.vi – 24.viii; record numbers: max. 68 on 10.vii, at sugar.

While field notebook records must be entered immediately, winter is the obvious time for routine paperwork, sorting and analysing the data while the year is still fresh in the mind. There are sure to be spells when the weather makes it impossible to look for pupae, unless you are as hardy as Reid of Pitcaple, who searched stone dykes in Aberdeenshire for cocoons of the Sweet Gale moth early one February. The snow was very deep and the cold intense 'which much interfered with a more successful hunt'. Nevertheless, he still found a dozen. We know this because he sent the information in to be published (Tutt, 1901–05). There are three separate elements in scientific recording. The first is the enjoyable one: fieldwork, making the observations. The second is the discipline of writing them up. The third and vital one is ensuring that the information is made available for use. Records can be published if of sufficient interest, or sent in to a data bank. It is a tragedy when a lifetime's work is destroyed because executors or distant relatives saw no value in a few old notebooks, and threw them on a bonfire. It is merely annoying when the latest distribution maps have blanks that could have been filled by one's own records. Ensure that they reach the right people.

Sending in records

From the early 1960s, the Biological Records Centre (BRC) at Monks Wood administered a national recording scheme for all the larger moths, based on the 10 km square unit. The data collected was the foundation for the 'dot maps' that have since been published for most of the macrolepidoptera except the Geometridae. Unfortunately, the scheme foundered in 1982 with the retirement of its guiding light, John Heath. In spite of a growing interest in moths, and the more widespread use of mercury vapour light traps over the next decade, there was no central repository for the records that were generated.

Thus, even for scarce and threatened species, a national overview of their current status was almost impossible. However, on behalf of the Joint Nature Conservation Committee, in 1991 Dr Paul Waring organised existing county recorders and recruited other active fieldworkers into a national network. Work was concentrated on macro-moths that were thought to occur in fewer than 100 (about 3 per cent) of the 10 km squares in Britain. Waring's enthusiasm galvanised the study and conservation of our scarcer species. Knowing that their records would be used proved a great stimulus to observers. It is to be hoped that this impetus will be maintained and extended. Computers now offer the means of exchanging and storing vast quantities of data far more cheaply and quickly than the old manual methods, so perhaps a resurrection of the Monks Wood BRC scheme is a possibility.

In the meantime, the county moth recorders collate all records for their area. Generally, the Watsonian vice-county boundaries are still used, ignoring modern administrative changes. Almost the whole of Britain is covered, although the system is very stretched in large parts of Scotland. Most of the recorders are volunteers, sometimes by default, and inevitably the level of organisation, efficiency and feedback varies. An up-dated directory is published at intervals (Waring, 1999), or other local observers can be asked to supply the relevant name and address. Additionally, some regions have a Biological Records Centre, often connected with a museum.

Usually, records are taken on trust imposing obligations on the sender. It is always reassuring to have an actual specimen or adequate photograph of anything out of the ordinary. Or, if the recorder lives nearby, he or she may be willing to verify unusual catches – few would resist an invitation to see a rare moth! New observers will need to establish their credibility, which is understandable. It is galling to be doubted, but a recorder's duty is to be scrupulously careful that local, and possibly national, data banks are not corrupted by inaccurate information. Do not send in a record unless you are absolutely certain of it. Equally, if you later realise it was an error (and we all make mistakes), withdraw it. This will enhance rather than diminish your reputation, proving you to be a careful and conscientious observer. (There is no need to admit being mistaken – merely say that you are no longer sure the record was correct.)

Appropriate records are also published in the annual summary of Lepidoptera immigrant to the British Isles, a feature that began as far back as 1931 and has continued ever since. Such long runs of data are especially valuable. There is also a microlepidoptera review of the year, listing new and unusual sightings, and all those that are the first for a vice-county. Records can be sent in via the county recorder, or direct to the compilers. Best of all, if you think they are of sufficient value, you could always write them up yourself and send them to a journal for publication.

Undertaking scientific projects

This may seem daunting at first, but anyone who is a good observer, and capable of committing their results to paper in a clear and logical way, should find it relatively straightforward. It is best to start with simple projects before moving on to more ambitious studies. The latter are likely to require some knowledge of scientific method, which might be thought to put the amateur at a disadvantage.

It is true that the professional scientist or academic has more resources to call on: laboratory facilities and equipment, technical support, easier access to reference libraries and journals, and an awareness of related research in the same field. However, being an amateur has its advantages too. It gives greater freedom to investigate topics of personal interest and pursue them for as long as is necessary, or abandon them if the results are disappointing. The professional researcher is far more restricted, and his choice of project often depends upon whether a grant can be obtained to pay for it. Normally this is for a limited period, at the end of which the outcome must be written up whether or not sufficient data have been obtained.

This pressure to produce results at any cost does not always make for good science. In Sussex, I was a voluntary warden at a nature reserve often used by researchers from the nearby university. Usually their projects were highly academic and of little or no practical value for conservation. One, however, seemed promising. Random sections of the centuries-old downland turf were removed, to record which plants colonised most successfully the newly exposed bare soil. Partway through the experiment, I noticed that a moat filled with metaldehyde pellets now surrounded each plot. Snails had been selectively grazing the tiny seedlings. When I naively suggested that the snails might be an important factor in the equation, the researcher soon put me straight. They could not be allowed to ruin the experiment by eating all the data.

For the amateur, most studies are not planned in advance, but happen almost by accident. Perhaps you develop an interest in a particular species, and gradually discover more and more about it. Almost before you realise it, your casual interest has evolved into a structured study. Soon you have more information about the species than anyone else, and some of this is new material not to be found in the books. You have the makings of a short note for the journals, or even a scientific paper. First it must be written up in the proper way.

Writing up the results

Decide which is the most appropriate publication for your contribution, as this will influence its style and tone. A relaxed and conversational approach is fine for a local moth-group newsletter, whereas a more formal and impersonal treatment would be appropriate for the scientific journals. Subscribe to the one (or more) that best suits you, and base your own contributions on its house style. Pay particular attention to the treatment of scientific names, dates and references.

It may be safest to start with short notes – simple accounts of unusual sightings or previously unrecorded behaviour (Plate 115). The species in question need not be rare: there are many gaps in our understanding of even the commonest moths. Emmet (1992) was unable to chart the overwintering stage of the Honeysuckle Moth *Ypsolopha dentella*, found in virtually every garden where that shrub grows. However, deciding whether a note or paper is worth submitting can be difficult. Does it add anything new to existing knowledge, or confirm or contradict previous studies? Might it be better to wait, and gather more data? Would the readership find the subject sufficiently interesting? Will the editor think it worth the space? It may be wise to consult him in advance: having a contribution rejected is always deflating. On the other hand, it is sad to think of all the valuable observations that must be wasted because they are never written up.

Having decided to go ahead, the first priority is to be absolutely sure of the identity of the species. This may seem obvious, but a researcher once spent months studying caterpillars of the Six-spot Burnet moth in the belief that they were those of the Chalk Hill Blue *Lysandra coridon* butterfly. He was remarkably unconcerned when I pointed this out. It did not really matter, he explained, as he could still use the data. All he needed to change was the title of his thesis.

Next, be concise, and avoid extraneous or irrelevant padding. A little circumstantial detail can put observations in context, and help to set the scene, but normally there is no need to tell readers what you had for

Plate 115: Buff-tip *Phalera bucephala* caterpillars entwine Medusa-like as they prepare to moult, reinforcing their warning coloration. This habit is not mentioned in the books. Behavioural observations such as this might merit further study followed by a short note to one of the scientific journals. (Banffshire).

breakfast or the make of camper van that took you to the site. Express your findings in a clear, straightforward way. English will not be the first language of some subscribers. Nor is semi-literacy the hallmark of a brilliant scientist, although a few would like to think so. Do not employ convoluted syntax and jargon in a misguided attempt to make your submission appear more erudite. Attempting my own first scientific paper, I included the phrase 'visual estimation of the numbers present'. My editor, Michael Shrubb, crossed this out and substituted 'counting'. I have been grateful to him ever since. So try to accept editorial amendments with good grace unless they distort what you meant to say. Part of an editor's job is to protect contributors from their own errors of carelessness, ambiguity or inexperience.

It is also easy to confuse fact with supposition. For instance, perhaps all the books say that a certain caterpillar feeds on a wide range of deciduous trees, but you are convinced that in your area it feeds only on Alder. This may well be correct, but it is an opinion, impossible to prove. What you really mean is that you have only ever found it on Alder, although you have worked many other types of tree equally thoroughly: that would be fact. All the better if you can provide actual counts as evidence. If your findings contradict those of previous authors, do not be too hasty in implying they were wrong. The results may be valid for your site, but perhaps the species behaves differently elsewhere. There are numerous regional variations in food plant, preferred habitat and emergence dates, for example.

Another danger is to assume that behaviour in captivity exactly matches that in the wild, or to generalise from a few observed instances. The Spinach caterpillars obviously mimic leaf stalks of their food plant, Blackcurrant (Plate 116, left). The pupa's resemblance to a bud, its skimpy cocoon being the white hairs near the base of the leaf stalks (Plate 116, right), also seems too much for coincidence, but is not mentioned in the literature. Although 12 of 14 captive caterpillars pupated in this manner, it would be safer to confirm the habit in the wild before reporting it to a journal.

Speculation in the absence of conclusive proof is a particular temptation for the amateur. If you must speculate, make it absolutely clear that you are doing so. (Proper scientists never speculate, for when they do it is described as 'putting forward a challenging hypothesis'.) On the other hand, show confidence in your own results if you want others to believe them, and do not continually hedge and prevaricate. We can all smile at the ultra-cautious scientist. If a UFO landed in his garden tomorrow, and the alien life-forms ate his hens, he would not accept what was happening until he had read an account of similar instances by three eminent colleagues, published in a reputable, properly refereed journal. Then he would submit his own short note, with several statistical proofs (one of them rather elegant) that the brown hens were eaten first.

Plate 116: The Spinach *Eulithis mellinata* caterpillars (left); pupa and cocoon (right). Captive stock, bred from a Banffshire female.

Analysing data

A basic knowledge of statistics is needed by amateurs too, enough to show whether or not data are significant. Mainly this is down to sample sizes: conclusions drawn from inadequate sets of figures are invalid in statistical terms because of the effects of chance. This does not mean that they are wrong, merely that better data are required to back them up. Even non-significant results may be sufficiently interesting to merit further investigation, or to suggest a line of enquiry that may grow into a methodical study. This should incorporate safeguards to reduce bias and ensure (as far as possible) that sampling is random. Seek academic advice if necessary. Normally, very sophisticated statistical analyses are unnecessary in amateur fieldwork, where there is ample scope to choose clear-cut projects likely to give unambiguous results. With practice it is easy to judge significance without doing the actual sums.

A broader perspective

Working a new site and discovering which moths are present is always exciting. Getting to know the same site intimately, and documenting the changes year after year, can be equally satisfying. It also promotes a more balanced view of the varying fortunes of our moths, how some species prosper and others decline. This acts as an antidote to the more alarmist outpourings of the conservation industry. Fluctuations in population and range are the norm, and always will be. Declines do not necessarily presage extinction. Looking back through old notebooks gives a reassuring sense of proportion. Sometimes it also shows the fallibility of human memory. The Clouded Magpie was abundant in my youth, or so I remember. Dozens could always be seen in the woods of the Irwell valley near my home. Digging out my logbook to confirm this recollection, I quickly found the relevant entry: 'abundant 1961, otherwise scarce'. That shows the value of recording.

Chapter 12

Conserving moths

Not all the factors that influence the fortunes of wildlife are under human control. Campaigning conservation pressure groups, in their endless quest for funds and influence, play down this point for obvious reasons. For every threatened species there must be an action plan, and the resources (money, basically) to put it into operation. In reality, we can do nothing about long-term climatic change, or the occasional disastrous spell of weather that causes populations to crash. It would be hard to influence the complex relationships between species – competition, predation and parasitism, disease – even if we understood them. The ramifications are potentially endless. Thus the White-spotted Pinion has declined because of a beetle that carried the fungus that destroyed its food plant, elm. Presumably, elm, fungus, beetle and moth have always co-existed, but something happened to disturb the equilibrium.

Fortunately, we are not completely helpless. It is within our power to aid a species by safeguarding its preferred habitat. Once its requirements are understood, existing sites can be managed in the optimum way, or rescued from deterioration. Perhaps former sites can be returned to their previous state, allowing the species to be re-introduced, or new sites created. Few readers of this book will have the opportunity to contribute, except financially, to such ambitious projects. However, most people do have a small area of potentially excellent habitat to manage as they wish. It is called a garden.

Gardening for moths

Changing attitudes

Increasingly, owners see their garden as a potential haven for wildlife. Such a viewpoint would have surprised previous generations – gardens then were a source of flowers, fruit and vegetables, and whatever might conceivably eat or damage this harvest was rigorously kept out or destroyed. In any case, there was plenty of wildlife in the countryside for those who liked it. In today's more affluent society we are less dependent on our own garden produce, and are more concerned that it should be grown organically, without chemical pesticides. Wildlife in the garden is seen as an asset to be encouraged, not a threat to be controlled, unless the amount of damage becomes intolerable. Supplying high-quality food for birds such as Blue Tits and Greenfinches is a substantial industry, and plants become popular just because they attract butterflies.

Welcome though these developments are, the main beneficiaries among birds are the commonest and most ubiquitous species already. Apart from House Sparrow and Starling perhaps, gardens are not the most important breeding habitat for any of the birds that regularly use them. Butterflies too are casual visitors, mainly for nectar from flowers such as Buddleia. Generally, gardens are not an important breeding habitat for this group either, except possibly the Holly Blue. It is ironic that the two other butterflies that do breed there regularly are not welcomed: they are the 'cabbage whites'. However, gardens are genuinely important for moths.

Moths of suburbia

Emmet (1992) charts the habitat preferences of every British moth, excluding adventives. Suburban habitats, defined as gardens, parks and the outside walls of buildings, are listed as the sole habitat in Britain for 32 species, and a principal habitat for 273 others. Thus over 300 moths are associated particularly with suburbia, although Emmet's choice of species is inevitably somewhat arbitrary. A further 300 moths are listed as having no obvious habitat preference. Many of these are familiar garden moths too, using this term to apply to the average suburban patch. Large mature rural gardens, surrounded by good natural habitat, will of course have a far wider range of species that cannot fairly be described as garden moths.

Benefits and drawbacks of suburban habitats

Food plant is the most obvious reason why some species are associated with gardens. Clearly, those that use horticultural plants which are not native to

Plate 117: Moths whose caterpillars can feed on a very wide range of plants often thrive in gardens. The Angle Shades *Phlogophora meticulosa*, with its distinctive resting posture, is a familiar urban moth. (Banffshire).

this country cannot survive anywhere else. The moths dependent on cypresses, Cotoneaster and Firethorn are good examples. A larger number have as food plants trees that, although native, are relatively local or scarce in the wild, but frequently planted in gardens and parks. These include apple, pear, plum and cherry, commonly grown both for ornament and fruit. Other moths benefit from trees often used for avenues, like poplars, limes and sycamores.

Additional advantages include the reduced number of specialist predators on moths. Warblers and bats are scarcer than in the open countryside. Mice and voles (which eat many pupae) are kept down by the artificially high population of one of their chief enemies, the cat. Towns and cities are significantly warmer, especially at night when most moths fly, and in winter, because of heat escaping from buildings. They are also more sheltered, with houses, fences and hedges acting as windbreaks. Thus the microclimate is kinder in towns, and this is particularly important for more southern species like many of the recent colonisers.

On the other hand, although gardens contain a greater diversity of plant species than almost any natural habitat, most of these plants are exotic, and therefore less suitable as food for indigenous moths. Even polyphagous caterpillars ignore many garden plants. Plantings are also widely fragmented and scattered, with few large single stands. Then there are the detrimental effects of gardening on moths, however unintentional: grass is mown unnaturally short, and rolled. The best tender young growth of shrubs is sheared off when hedges are trimmed. Soil where caterpillars and pupae are hiding is dug, raked and hoed. Native food plants useful to many species are removed by weeding for the benefit of the useless exotics. Once flowering is over, stems of herbaceous plants are cut and burnt before they seed, destroying a potential source of food and shelter. Air pollution from industry, domestic fuel and exhaust gases is still a factor, though not as bad as it was. And not all gardeners forego insecticides. Should a moth still manage to complete its metamorphosis, it risks being distracted at night by artificial light.

Managing a garden for moths

Even the most ordinary garden will already support moths, especially if it is slightly neglected, weedy and overgrown. Therefore, encouraging moths does not necessarily involve extra work. Rather the opposite: it is a good excuse to be less tidy, and more relaxed. There is no need to go to extremes, and allow the plot to return to wasteland – that would be counter-productive anyway, supporting fewer and less interesting species than a properly managed garden.

Nevertheless, potentially harmful operations should be reduced, or their effects minimised. Hedges must be trimmed, or else they will not remain hedges, but it is possible to do this less often. If the fallen clippings can be left to lie for a couple of days before disposal, this will give caterpillars on them the chance to climb back into the hedge. Hoeing between herbaceous perennials and around the foot of trees and shrubs, potentially damaging to pupae, can be avoided by using ground cover plants instead. Many, like Yellow Archangel, will serve as food plants too, while Bugle makes a good nectar plant as well. In autumn, do not be too eager to remove and burn dead stems and shrivelled leaves and flowerheads in which eggs or small caterpillars might be hibernating.

Needless to say, insecticides and other sprays should never be used unless essential, and then selectively. Few garden moths are harmful, though several can be minor nuisances. Even the keenest lepidopterist would not want the Codling Moth *Cydia pomonella* to damage too many apples, and the Pea Moth *C. nigricana* caterpillar is an unwelcome discovery when pods are being shelled. (I would be pleased to find either in my Banffshire garden, however, since they would be new for the vice-county.) Otherwise, perhaps only the caterpillar of the Cabbage Moth *Mamestra brassicae*, boring into the heart of a cabbage and fouling it with frass, regularly stretches tolerance beyond breaking point. A healthy bird population normally prevents other caterpillars becoming sufficiently numerous to cause damage.

Any garden can be made more appealing to moths by introducing particularly favoured plants. Many are already widely grown, being useful and attractive in their own right. They can be divided into food plants and nectar sources, although some double as both.

Food plants for moths

It seems logical to provide most help to those moths that are restricted to gardens, or at least far commoner there than in the wider countryside. For tall hedging or ornamental conifers, cypresses are the obvious choice, with the Monterey Cypress surely preferable (where frosts permit) to the dreaded Leyland. Almost anywhere in England, Blair's Shoulder-knot (Plate 118) will quickly take advantage, and within their ranges the Cypress Carpet, Cypress Pug (Plate 48) and Freyer's Pug might appear. Lawson's Cypress is also good, and junipers can add a few more species including the Juniper Pug and the recently arrived micro *Argyresthia trifasciata*. However, do not despair

Plate 118: The popularity of cypress hedges has helped the rapid colonisation of southern Britain since 1951 by Blair's Shoulder-knot *Lithophane leautieri*, originally a Mediterranean species. (Sussex).

Plate 119: The Phoenix *Eulithis prunata*, although found very sparingly in woodland, is predominantly associated with gardens and allotments where currants are grown. (Sussex).

if your existing hedge is privet, as it will be used by several of the Ennominae including the Scalloped Hazel and Willow Beauty, plus the Waved Umber *Menophra abruptaria* in some areas. The Privet Hawk-moth (Plate 27) is a possibility in southern England, at least in theory, but during 25 years in Sussex I failed to find the caterpillar on this shrub.

Fruit trees and bushes support several predominantly garden moths. The caterpillar of the Green Pug has a most refined diet, eating only the blossom of apple and pear. I have never seen this moth away from gardens. A much rarer apple feeder is the Red-belted Clearwing *Synanthedon myopaeformis*, and any tree that supports its wood-boring caterpillar deserves special protection. The related Currant Clearwing is one of several moths associated with currant and gooseberry bushes. Although still listed as an injurious pest in the horticultural literature, today it seems to be scarce and declining. I was delighted to have a colony in my Brighton garden, and spent hours watching the adults buzzing like small black wasps around the bushes, or sunning themselves on the leaves. Admittedly, some of the fruit was lost when overloaded stems snapped off at the emergence holes, but I regarded the moths as part of my crop too. It was a bumper year when three mated pairs were in view at once. Other attractive currant specialists include the Magpie Moth, V-moth (Plate 51), the Phoenix (Plate 119) and the Spinach. Little round holes in the lower leaves are a sure sign of the Currant Pug caterpillar at work.

Raspberries are also worth growing. They support the Fan-foot *Zanclognatha tarsipennalis*, whose caterpillar feeds on the withered leaves, a strange diet. If there are wild ones nearby the lovely Peach Blossom (Plate 81) and its remarkable caterpillar (Plate 41) may be present. Surprisingly, the inconspicuous Raspberry flowers are a favourite nectar source for noctuids at dusk, yet butterflies show not the slightest interest (although they like the related bramble). As with currants, in July the berries will benefit birds, especially Whitethroats and juvenile Blackbirds. Usually they leave enough for jam.

Plate 120: The Golden Plusia *Polychrysia moneta*, gleaming like a newly minted coin, lives up to both its English and scientific names. Lacking food plants in the wild, herbaceous borders are its sole habitat in Britain. (Sussex).

The moth enthusiast's herbaceous border is incomplete without a few clumps of Delphinium and Larkspur for the aptly named Golden Plusia (Plate 120). Although widespread through most of Britain since its colonisation over a century ago, non-native food plants make it entirely dependent on gardens. Sadly, some growers regard it as a nuisance – the 'delphinium moth' (Baker, 1983). The caterpillar bites through the main ribs of a leaf, causing it to collapse like a broken umbrella, but the flower spikes are unaffected. Every year I used to count the bulbous yellow cocoons spun under the lower leaves of one clump in my own garden. The record was 12, yet even then the plant showed only minor feeding damage. Such a concentration inevitably attracted predators, in this case ichnuemons. A tiny round hole in almost every cocoon showed where an ovipositor had pierced it. Next year there were no Golden Plusia caterpillars on the clump, a fine example of density-dependent regulating factors in action.

In south-east England, a moth garden should contain Sweet William for another attractive recent colonist, the Varied Coronet. Whether or not this deigns to use it, the commoner *Hadena* species certainly will. Wild campions are worth including for this reason too, and for their nectar, which is particularly attractive. Otherwise, the choice of plants will depend partly on soil type, the neighbouring habitat, and the local moth fauna. As a group, the pugs seem particularly well-suited to gardens; being small moths perhaps they do not need large stands of food plant to support a viable population. Just two clumps of Bladder Campion, self-sown beside a box hedge in my chalk garden, maintained a colony of the Netted Pug *Eupithecia venosata* for many years. In the same garden, the pretty little Toadflax Pug (Plate 96) thrived on exotic Purple Toadflax and probably cultivated Snapdragon, being much more numerous there than on nearby downland where its normal food plant, Common Toadflax, grew.

If the garden is large enough, it is fashionable to designate part of it (usually the furthest from the house) as a wild area where weeds and native plants are allowed to flourish. This brings obvious benefits for moths, albeit mainly common and ubiquitous species not particularly associated with gardens, or in much need of help. Nettle beds ostensibly created for the sake of Small Tortoiseshell and Peacock (Dunbar, 1993) are unlikely to attract those butterflies, which are extremely choosy when egg laying, but are almost certain to support moths such as the Burnished Brass, the Snout *Hypena proboscidalis* and the Mother of Pearl *Pleuroptya ruralis*. Where bindweed cannot be eradicated, at least the delicate White Plume Moth *Pterophorus pentadactyla* can be expected. In acid soils, Foxgloves readily naturalise, enticing yet another *Eupithecia*, the Foxglove Pug, into the garden. At least it is one of the more distinctive of the genus.

Nectar flowers

Whilst most of the flowers that attract butterflies are also good for moths, the reverse is not necessarily true. The latter use a wider range of nectar sources, especially those that release most of their fragrance in the evening to attract moths as pollinators. Various lilies, Evening Primrose, Night-scented Stock, Sweet Rocket and White Jasmine are in this category. Wandering round a garden filled with such plants at dusk is a delightful experience: many of the blossoms seem almost luminous in the half-light, and the heady cocktail of different scents is intoxicating.

Honeysuckle in full bloom can be cloyingly sweet on a still, balmy evening. Because of the length of its tubular corolla, few moths have a proboscis long enough to reach the nectary. Hawk-moths have no such difficulty, and it is marvellous to watch the speed and precision of their hovering as they sip at each flower in turn. Only a few noctuids are equipped to copy them, particularly the Shark Moth and, in the north, the Gold Spangle *Autographa bractea* (Plate 111). Nor should the value of honeysuckle as a food plant be overlooked. The Early Grey *Xylocampa areola* and the Lilac Beauty (Plate 19) follow it into gardens, as

Plate 121: The Twenty-plume Moth *Alucita hexadactyla*, a member of a small family with wings divided into six feathery lobes, occurs in almost every garden where honeysuckle grows. (Banffshire).

do two micros, both distinctive enough to have acquired vernacular names: the Honeysuckle Moth and the unique Twenty-plume Moth *Alucita hexadactyla*, with feathers rather than wings (Plate 121). Later, the berries are a favourite food of wintering Blackcaps.

Buddleia, so pre-eminent for butterflies by day, is equally good for moths after dark, or even in the daytime. Ideally, the bush should be in full view from the kitchen window – a Humming-bird Hawk-moth makes a welcome diversion during the washing-up, although Silver Ys are more usual. Because the flowers are small, even geometrids and pyralids can reach the nectar. In Sussex, I used to find the Vestal there in good migrant years, well set off against the deep royal purple. Although using both sugar and a light trap, every year there are a few moths I see only at Buddleia, especially the Dotted Rustic. If there is room for more than one bush, try to stagger their flowering times to extend the season: some individual plants bloom earlier than others. Judicious pruning will also help.

Perhaps the ultimate example of a nectar source often grown specifically to attract moths – or rather, one particular moth – is Nicotiana or the Tobacco Plant (tall, old-fashioned varieties are best, as some of the new cultivars seem to lack scent). Traditionally this is a favourite of the Convolvulus Hawk-moth, a huge and spectacular migrant (Plate 122). It is scarce enough to attract interest and excitement whenever it appears, but not so rare that hopes of enticing one to even the smallest garden, anywhere between the South Coast and Shetland, would be unrealistic. Year after year I grew Nicotiana prolifically without result. It seeded itself everywhere, smothering

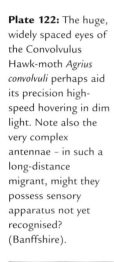

Plate 122: The huge, widely spaced eyes of the Convolvulus Hawk-moth *Agrius convolvuli* perhaps aid its precision high-speed hovering in dim light. Note also the very complex antennae – in such a long-distance migrant, might they possess sensory apparatus not yet recognised? (Banffshire).

the onions and competing with the carrots. Ironically, when a Convolvulus Hawk did at last pay a visit, I mistook it for a bat at first. It was worth the wait. I had expected such a big moth to be ponderous and clumsy, not so swift and agile that in the gloaming it seemed to vanish and re-materialise elsewhere rather than move from flower to flower. Attracting this moth is pure self-indulgence. There is no conservation value since it hardly ever breeds here, but only a puritan would cavil at such harmless pleasure.

Conservation in the countryside

Why conserve moths?

Perhaps only affluent western societies can afford to pour resources into conserving species of no obvious value for food, medicines, clothing, tourism or any other conceivable purpose. Conservation for its own sake, or from a sense of moral obligation, seems a luxury. While it would be cynical to deny altruism as one motivating factor, perhaps the urge to conserve threatened species, however obscure and unattractive, is not entirely free from self-interest. We share a considerable part of our DNA with all other species on this planet, even moths. Secondly, since we no longer need to spend every waking moment in the quest for food, warmth, shelter, safety, mates and territory, something else is needed to occupy our leisure time. Rather than being mere diversions, hobbies and interests must increasingly provide the mental and physical stimulation vital to health and personal fulfilment. Wildlife fills the void for many. Ensuring that a rich and varied fauna and flora survive to be studied and enjoyed may not be such a luxury for advanced societies after all.

But is conservation 'unnatural'?

Even conservationists disagree about how much interference in the fate of our wildlife is justified. At one extreme, there is the view that we should act merely as neutral observers, documenting changing fortunes without deliberately trying to manipulate them. This is a valid approach in regions that have vast areas of pristine wilderness where humans have had little direct impact on the natural ecosystems. It would be best to keep it that way.

However, in a densely populated country such as Britain, human influence on the environment is already all-pervasive. At best, only a few areas of semi-natural habitat survive, and even these may be affected by air pollution, fragmentation and the influence of the surrounding altered land. Humans have already had an enormous impact here, and will continue to do so. Much (but not all) of this impact has been detrimental to wildlife.

There is a good case for attempting to redress the balance where possible, by taking positive measures to help species that have become threatened mainly as a result of human actions.

Even so, there can be debate about how much aid is justified before the exercise becomes too artificial, akin to maintaining a feral rather than a truly wild population. The re-introduced Large Copper butterfly at Woodwalton Fen survived only through regular topping up with captive-bred stock. This is an extreme example, but many other threatened plants and animals depend for their continued existence on some degree of human tolerance or help. Is that 'unnatural'? Here, perhaps the argument becomes philosophical. If *Homo sapiens* is regarded merely as a species of mammal, albeit a highly evolved and successful one, then we are an integral part of most ecosystems of this planet. It can be argued that anything humans do, deliberately or incidentally, or choose not to do, is perfectly valid behaviour that is characteristic of the species in question. Is conserving moths like the Kentish Glory for our aesthetic enjoyment wholly dissimilar from ants guarding Chalk Hill Blue caterpillars for the sake of their sweet secretions? If the introduction of fish to new waters as eggs stuck to the feet of wildfowl is regarded as a 'natural' means of dispersal, why should the inadvertent transportation of species including moths by human agency, in the hold of a ship or aeroplane, be regarded as fundamentally different? Obviously, the fact of human involvement should be recorded, if known, as should any inter-relationship between species.

Withholding aid

However, even the most fervent conservationist accepts that some species are more deserving of help than others. The Human Flea and the Bed Bug are among our scarcest and most threatened insects, yet tramps are not designated as Sites of Special Scientific Interest to safeguard their habitat. Some selectivity is also attached to the conservation of moths. Micros as a whole receive little attention, even when they are not regarded as harmful. Several members of the Tineinae were all too common inside houses during my own youth, doing enormous damage to grandmother's stored fabrics, furs and feathers. Today they are rapidly declining. The Tapestry Moth *Trichophaga tapetzella*, once an abundant and troublesome pest, is now almost extinct in Britain. The Common Clothes Moth *Tineola bisselliella*, another formerly ubiquitous villain, has been declining for half a century and is now local and scarce. I have not seen one for many years, although several of my jackets still bear feeding scars. Its relative the Case-bearing Clothes Moth *Tinea pellionella* shows a similar decline, now surviving mainly in unheated outbuildings and birds' nests in roof spaces. Artificial fibres, improved hygiene, and particularly the dry atmosphere created by central heating, are the factors responsible. As yet there is no campaign to improve the fortunes of these destructive moths. Likewise, when nearly 400 Gypsy Moth caterpillars were found in Essex gardens in 1995, the return of this long-lost British species was not celebrated. Instead, all the caterpillars and eggs were methodically destroyed by the Forest Authority to forestall potential damage to woodlands (Waring, 1995b). Such caution is inevitable, although some moths that are forestry pests on the Continent are harmless here, like the Black Arches (Plate 123).

The history of moth conservation in Britain

Butterflies such as the Large Copper, Large Blue and Purple Emperor *Apatura iris* have long been afforded a degree of help in Britain, although this has not always been successful. Nature reserves have a particularly poor record in this respect. Monks Wood, for example, once renowned for its butterflies, has

Plate 123: Sometimes a major pest of conifer plantations elsewhere in Europe, in Britain the Black Arches *Lymantria monacha* is a blameless species of southern oakwoods. (Sussex).

lost at least eight of its more interesting species since it became a reserve (Pollard and Yates, 1993). However, positive action aimed specifically at moths is a relatively recent development, although concern for this group is not. Already by the second half of the nineteenth century, collectors and dealers were being urged not to take too many specimens of certain scarce and local moths, which at that time included the Brown-tail. The Royal Entomological Society set up a Protection Committee in 1923, which was enlarged with representatives from other interested bodies to become the Joint Committee for the Conservation of British Insects (JCCBI) in 1968. Mainly, these committees tried to protect threatened species by regulating the excesses of collectors, whereas the modern approach centres on conserving their habitat. Although sites for rarities have been deliberately kept quiet and informally protected in the past, the first reserve declared specifically for a moth may well have been Vert Wood in East Sussex in 1951. This unsuccessful attempt to save the Lewes Wave from extinction has already been mentioned.

Legal protection for a small and somewhat arbitrary selection of moths, a mere five species, followed in 1981 under Section 5 of the Wildlife and Countryside Act. Reviews have seen the list extended only slightly, with the Fiery Clearwing *Bembecia chrysidiformis* and Fisher's Estuarine Moth *Gortyna borelli* the latest additions in 1998 (Waring, 1998). If anything, this first statutory attempt to help Britain's most threatened moths simply highlighted our lack of knowledge. One species on the original list, the Viper's Bugloss, was already extinct, and has since been removed. The then Nature Conservancy Council began the National Moth Conservation Project in 1987 to promote the study and recording of those moths protected by law, and other species that were thought to be rare from the piecemeal and often out-of-date information available. A review of all Britain's moths, based on objective criteria, was clearly long overdue.

Building on the work of Hadley (1984), the Red Data Book for insects (Shirt, 1987) attempted to classify the status of the scarcer macro-moths according to the number of 10 km squares from which they had been recorded. Strictly speaking, this is a measure of range and distribution rather than abundance, nor is a species necessarily threatened if it is limited to a very few sites because of unusual habitat requirements. Several of the Scottish montane moths fall into this category, as do some coastal ones. Conversely, a species can be widespread, but still highly vulnerable. Nevertheless, the correlation between restricted distribution and the need for conservation measures is generally sound: obviously, the fewer sites that support a species, the more necessary it is to safeguard them all if possible. Shirt's list has since been revised and amended, as fieldwork has shown many of the moths on it to be more widespread than previously thought, and others less so.

The Earth Summit in Rio de Janeiro in 1992 stimulated an interest in active conservation measures under the banner of biodiversity. This was a relatively new concept, urging the protection of all species, however humble, for their own sake. As a result, in 1994, the UK Biodiversity Action Plan was launched by the government. Although heavily biased towards birds and butterflies, a Steering Group Report in 1995 included a welcome number of moths as meriting aid. In order of decreasing priority these were graded as Short List (3), Middle List (50) and Long List (69) species (Bourn and Warren, 1997). Somewhat confusingly, the moths chosen were not necessarily those with Red Data Book status, nor were all of them obviously at risk compared with others that had been omitted. This was in turn superseded by a new series of Action Plans for 53 Priority Moth Species (Bourn *et al.*, 1999). The sole micro was *Coleophora tricolor*, known only from the Norfolk Breck and Switzerland. Butterfly Conservation was appointed 'Lead Partner for Lepidoptera'.

In reality, the requirements of many of these moths are still very imperfectly understood. Thus the rather formulaic Species Action Plans combine calls for ecological studies with ambitious statements of intent. While it is good to see moths getting the attention they deserve, the volume of jargon and layers of bureaucracy engendered (though perhaps inevitable) cause many hearts to sink. We can only hope that most of the budget is not consumed by paperwork and salaries, and that moths themselves will actually benefit.

Choosing which moths to help

It is not practical to help all our threatened moths. Funds are always inadequate. Some local conservation groups achieve wonders with voluntary labour, but this too is a scarce resource. Inevitably, help can only be given to a few of the moths that apparently need it, so it is better to concentrate efforts where they will do most good. First, we must know why a species is declining. Even if unlimited resources were available, little can be done about adverse climatic change, inbreeding, or unsupportable levels of competition, predation and parasitism. Often these factors are hard to distinguish from deterioration of the habitat; indeed, if their habitat were ideal the species might easily be able to withstand them. So, in practice, conservation is overwhelmingly about habitat management and protection. It is no coincidence that so many of the Priority Species occupy transitional habitats that inevitably become unsuitable through vegetation succession unless maintained by exactly the right amount of grazing, cutting or burning.

Sometimes the requirements of a species are patently obvious, especially if it is possible to compare sites where it is thriving with those where it has declined. However, it is easy to overlook features that are important only at particular stages in the life cycle or certain periods of the year. Accordingly, the scientist will call for a programme of methodical and objective fieldwork, and there are times when such an approach is invaluable. Equally, there is a case for relying on the instincts of local observers with a long and intimate knowledge of the species. All naturalists must have had the feeling 'this spot looks absolutely perfect for ...' and been rewarded, too often for coincidence, by a sighting of the bird, mammal, insect or plant in question. It is satisfying to do, and then to boast about having done. Beside a deep pool on the River Deveron one April, I confidently began searching behind loose bark for pupae of the Small Clouded Brindle. It was not on the Banffshire list, and the distribution map showed no localities so far north on the British mainland. I had last found pupae of this species over 30 years earlier, by a stagnant, polluted, derelict canal in industrial Lancashire – a far cry from a fast-flowing Scottish salmon river. Yet eventually my search was successful. Why the site looked, felt or smelled right is impossible to explain.

Even if we are fairly sure we know what a species requires it may not be practical to help it. The Light Crimson Underwing *Catocala promissa* and the Dark Crimson Underwing *C. sponsa* are Priority List species with contracting ranges, evidently in serious decline. They are handsome moths whose extinction would be a major loss. Unfortunately, both are found only in large, unfragmented areas of mature oak woodland, where they normally occur at a low density. Creating suitable new habitat in overcrowded southern England is hardly feasible, and would take centuries. The best we can do is to try and protect existing sites.

A slightly different problem arises with the Sword-grass (Plate 101), now found regularly in Britain only in north-east Scotland, and considered scarce and local elsewhere in Europe too. Because it breeds on my own land, I have taken a particular interest in this moth for the past decade. The female lays a huge number of rather small eggs, given her size, suggesting a high mortality in the early stages. Food plants are thought to be grasses at first, followed by various common low plants such as sorrel and dock. Moths emerge in late September or early October, then hibernate until March. Overwintering survival seems to be good: moths are almost as numerous in spring as in autumn. Adults disperse widely: of 43 marked individuals released in September 1996, not one was ever seen again. Otherwise, its precise requirements are obscure, assuming that it has any. There is nothing exceptional about the habitat where it is found. In north-east Scotland it seems to occur at relatively low density over rather ordinary mixed farmland that does not support any other unusual species. I could not begin to suggest measures that might help this moth. Nevertheless, it is a Priority List species, so an Action Plan must be written!

It is very much easier to help moths that live in small, discrete colonies, especially if they depend upon a single food plant. Fieldwork to ascertain the moth's needs is then greatly simplified. If only a few hectares of land are involved, maintaining and improving the habitat is a realistic proposition, and so is monitoring of the results. The site can be purchased and designated as a reserve, or at least made subject to a management agreement. To date, all the reasonably successful projects to conserve moths have involved localised colonial species rather than those with dispersed, low-density populations like the Bordered Gothic *Heliophobus reticulata* or Buttoned Snout *Hypena rostralis*.

Back from the brink

The New Forest Burnet (Plate 124) was thought lost to Britain when the last colony in the New Forest became extinct in 1927. There was surprise and delight when a new population was discovered in Argyll in 1963, but this turned to dismay after overgrazing by sheep seemed to have eliminated the species from Britain for a second time by the mid-1980s. Fieldwork sponsored by the World Wildlife Fund in 1990 showed that it had hung on, surviving on tiny ungrazed cliff ledges, although the population was perilously small. Fences were erected to exclude the sheep, and the site was monitored annually under the auspices of Scottish Natural Heritage, but numbers remained low. Young (1997) was uncharacteristically pessimistic about its chances of survival, since only four individuals were seen during 10 days of fieldwork at the site in 1996, after rock falls breached the fence and led to severe over-grazing.

Fortunately, since then the moth has prospered. After 4 suc-

Plate 124: Rescued (for the present) in the nick of time, the New Forest Burnet *Zygaena viciae* will always be vulnerable at its only known British site. (West Argyll).

cessful years, the estimated population of the colony exceeded 340 adults in 1999 rising to over 1,000 in summer 2000 (DA Barbour, personal communication). Clearly this species is adapted to survive severe crises. Like others in the family, caterpillars may go into diapause a variable number of times, especially when conditions are unfavourable, taking 2 or more years to complete their development. Tremewan (1985) discusses the probable advantages. Almost certainly this is what saved the Argyll colony in 1996, together with the measures put in place to restore its habitat. In the short term at least, this has been a conservation success story, if at considerable (but worthwhile) effort and expense.

Re-introductions

Almost everyone agrees that introducing a foreign species is unwise. There are numerous instances where it has caused unforeseen harm to indigenous wildlife, compared with very few where the introduction proved an unqualified asset. The delightful Little Owl evidently filled a vacant niche in England and Wales, and if Allan (1948) was correct, the Jersey Tiger *Euplagia quadripunctaria* also turned out to be a welcome addition in Devon. These are exceptions, and anyway the practice is now illegal in Britain.

Opinions are split on the desirability of re-introducing species that have become extinct. Some naturalists remain steadfastly opposed to such meddling, but the general view is increasingly favourable, especially if the extinction was caused by human agency in the first place. Few ornithologists who originally opposed the re-establishments of the Red Kite in England and Scotland, or the Sea Eagle in the Hebrides, would now wish them to fail.

Nevertheless, most would have preferred it had they returned spontaneously, like the Osprey. Quite possibly they would have done so, but now we shall never know.

Often, attitudes are coloured by personal experiences. Observers in the south of England who have seen many precious colonies of scarce species destroyed by agriculture, roads and housing tend to be more in favour of compensatory action than those who live in relatively unspoilt parts of the country. There can still be debate about how far this action should go. All naturalists agree that existing habitats and communities of species should be protected. Most also approve when habitats that have deteriorated are restored to their former state, making them suitable once again for species that have died out. The dilemma is whether to re-introduce those species, or simply wait, in the hope that they will return of their own accord.

Where moths are concerned, there is every possibility that this will happen – eventually. Moths are mobile: as records on oil rigs, light ships and remote islands show, even non-migratory species disperse or stray. If habitat and climate are suitable, sooner or later the site will be recolonised if it is within a species' range. Unfortunately, this might take 50 or 100 years – not long in evolutionary terms, but perhaps beyond the observer's lifetime. It is tempting, then, to try and hasten the process, especially if the moth in question is known to be relatively sedentary. Candidates for re-introduction might include the Kentish Glory to its former English sites, now that its requirements (regenerating birch saplings) are known (Barbour and Young, 1993). The return of this fine moth would give pleasure to many, but it is unlikely to happen without assistance as the female is rather sluggish, and can barely get airborne with a full complement of eggs.

The Scarlet Tiger is another very local, highly sedentary moth that evidently has difficulty in regaining lost ground. Yet colonies originally established to provide material for genetic research have thrived (Kettlewell, 1973), including one in Cheshire, outside its British range, founded in 1961 and still extant (Rutherford, 1994). It is surprising that attempts have not been made to re-introduce such an attractive, day-flying moth to its former sites in East Anglia.

After this, the list of potential re-establishments could include every species we have lost – the Viper's Bugloss, perhaps, or the Spotted Sulphur. Both are attractive and distinctive moths, unlikely to return soon without help. If resources were unlimited, and work was properly documented and monitored, a case could be made out for such interventions. Otherwise, most naturalists would prefer to see efforts and expenditure concentrated on those threatened species that still survive. If re-introduction became standard practice, there is a danger that the loss of colonies would then be regarded as less of a tragedy, undermining the imperative to prevent it happening in the first place.

The European perspective

Often, species that are scarce in Britain because they are at the edge of their range are widespread and common over the rest of Europe. Conversely, some that we take for granted have surprisingly limited distributions abroad. Woods carpeted with Bluebells, to us so redolent of springtime, are a quintessentially British phenomenon, and gorse too has a predominantly Atlantic distribution. Hawthorn scrub is a scarce habitat elsewhere in Europe. Among birds, we are aware of our responsibility for the Great Skua as Britain holds 58 per cent of the Northern Hemisphere population, and for the Gannet, with 70 per cent of its world population in the British Isles.

Estimates of total population size are not available for moths, and in any case figures would often vary enormously from year to year. However, Karsholt and Razowski (1996) chart the distribution of all European Lepidoptera, dividing the continent into 36 'countries' largely corresponding with present political boundaries, which at least gives an indication of range.

It is slightly disconcerting to find that so many of the moths considered of high conservation priority in Britain occur almost throughout Europe. What we parochially call the New Forest Burnet is in fact found in 27 other countries from Norway to Greece; our Kentish Glory in 24 others. Most of the East Anglian specialities are equally widespread. For the following species, the number of other European countries in which they are found is given in brackets: Tawny Wave *Scopula rubiginata* (28) (Plate 125), Grey Carpet (20), Small Dotted Footman (20), Marbled Clover (30). Nor do rarities from other English regions fare differently: Bright Wave *Idaea ochrata* (27), Barberry Carpet *Pareulype berberata* (20), Speckled Footman (28), Striped Lychnis *Shargacucullia lychnitis* (25) and the Four-spotted (32). Likewise, moths we consider to be as Scottish as haggis and bagpipes are far from restricted to that country: Rannoch Looper (21), Scotch Annulet *Gnophos obfuscatus* (20), Cousin German *Protolampra sobrina* (21), Rannoch Sprawler (23) and Sweet Gale (31). Admittedly, the truly montane species are slightly more restricted by habitat: Mountain Burnet (13), Netted Mountain Moth (14), Black Mountain Moth (13), Northern Dart (16) and Broad-bordered White Underwing (10).

Plate 125: The Tawny Wave *Scopula rubiginata* is a most unusual colour (when fresh) for a member of its group. One of the Breck's rarities, it has benefited from conservation headlands around fields. (Norfolk).

This is not an argument for reducing efforts to safeguard these species in Britain. The Osprey and the Barn Owl have almost cosmopolitan distributions, yet rightly much effort is spent on their protection here. If the Kentish Glory became extinct in Scotland, it would console me only slightly to know that I could still see one by travelling to Norway or France. If the Sword-grass died out on my own land, I would miss watching it at sugar every autumn, and marking spring by its reappearance from hibernation.

However, the incentive to conserve species is undeniably greater if they are scarce or localised elsewhere. Some moths considered in need of protection in Britain are certainly in this category, being found in fewer than 10 other European countries. They include Fenn's Wainscot *Chortodes brevilinea* (9), the Sandhill Rustic *Luperina nickerlii* (8), and the Marsh Mallow Moth *Hydraecia osseola* (6), all very local everywhere. The Rannoch Brindled Beauty (6) is unusual in being entirely boreal – unlike most other Arctic species it is absent from the Alps. Of our two macro-moths associated with Sea Wormwood, we have lost the Essex Emerald. In a sense, it was the less precious because it occurs in 25 other European countries, whereas the species that shared its habitat and food plant is much more restricted abroad. This is the Scarce Pug *Eupithecia extensaria* (4), which otherwise has an eastern distribution, being found in Latvia, Lithuania, Romania and European Russia. Perhaps it was introduced to England by trading vessels, long ago.

Some of our moths, though widespread in Europe, have evolved into distinct geographical races here. Rather arbitrarily, a few races have been raised to the status of subspecies. Others, equally distinct if not more so, have not. The British form of the Northern Dart, named subspecies *alpina* (Plate 71), well deserves that status. It is a most attractive moth, richly coloured and strongly marked. The nominate subspecies is Scandinavian. Specimens sent to me from Finland were smaller, duller and plainer than ours. There, the moth occupies a different habitat, living in woodland rather than on bare mountain summits (G Varkonyi, personal communication). The bluish-grey race of Ashworth's Rustic found in North Wales (Plate 126) is also very different from paler Continental forms, but in this case it is the nominate subspecies as the moth was first described from Welsh specimens.

All such races are interesting, as they illustrate evolution in action. Their extinction would mean the loss of a small, but unique, element of our fauna, whether or not the species itself survives elsewhere. Unfortunately, some have already perished. The Juniper Pug is locally abundant in many parts of Britain, especially in the north. However, the very pale form, subspecies *anglicata*, that occurred on the chalk cliffs of Dover has not been seen since 1915. The Welsh subspecies *segontii* of the Transparent Burnet is likewise feared extinct, as it was last recorded in 1962.

Plate 126: The slate blue nominate race of Ashworth's Rustic from North Wales differs strongly from the paler and more lightly marked forms found elsewhere in Europe. (North Wales).

Even more important are the endemics – species found nowhere else. Britain has few, but among micros there is *Eudarcia richardsoni*, only known from the Dorset coast. Of the macros, the status of the Northern Footman has been downgraded to a form of the Scarce Footman *Eilema complana*. The Arran Carpet *Chloroclysta concinnata* has a slightly better claim to be regarded as a good species, rather than a univoltine race of the Common Marbled Carpet. The latter is single-brooded over much of Scotland, including my own area, and very variable. However, I have never seen examples approaching those figured for the Arran Carpet in Skinner (1998) or Brooks (1991), especially as regards the clear and unbroken white submarginal line on the forewings. That leaves the Small Autumnal Moth *Epirrita filigrammaria* (Plate 127), a common moorland species found north-west of a line from the Severn to the Wash, as our only generally accepted endemic macro-moth. Although on wing pattern it is impossible to separate it from the closely related Autumnal Moth *E. autumnata*, the caterpillar has a distinguishing yellow subdorsal line. Both occur on my own site, and certainly behave as different species. The Small Autumnal flies from around 20th August on the heather moor, and has been over for at least a month before the first Autumnal Moths emerge in mid-October amongst birches.

Plate 127: The Small Autumnal Moth *Epirrita filigrammaria* is believed to be a purely British species. Fortunately it is not threatened, nor is it the subject of any conservation action. (Banffshire).

Finally, there are what might be called the unappreciated moths – species that are not regarded here as particularly special, but which have rather restricted distributions on the Continent. Again, the number of European countries besides Britain and Ireland in which they are found is given in brackets. They include the Mottled Grey *Colostygia multistrigaria* (8), Early Moth (9), Feathered Ranunculus (9), Lunar Underwing (8) and, remarkably, the Anomalous *Stilbia anomala* (5). None of these is classed as a Notable moth in Britain, or is thought in need of an Action Plan.

Incidental conservation of moths – the piggy-back factor

Moths will always be a minority interest, even among naturalists. Fortunately, measures to help wildlife with greater popular appeal are almost certain to help less favoured groups as well, especially where they involve habitat protection. Conservationists themselves take advantage of this. It is standard practice to raise support by publicising the plight of a high-profile 'flagship' species, knowing that the spin-off effects will benefit the many others that share the same environment but are less marketable. Thus in 1993, Butterfly Conservation acquired Catfield Fen in Norfolk primarily for the glamorous Swallowtail butterfly, but this land also supports wetland moths such as the Reed Leopard *Phragmataecia castaneae*, Small Dotted Footman (Plate 47), and Fenn's Wainscot. Though drabber, these are equally rare in Britain. In a European context, the last moth in particular merits more protection than the Swallowtail, given its rather restricted distribution abroad.

The Royal Society for the Protection of Birds (RSPB), with over a million members and more than 120 reserves, indirectly helps to conserve numerous rare moths. In the Scottish Highlands, Abernethy Forest and Insh Marshes are home to the Rannoch Sprawler as well as the Crested Tit and Spotted Crake. Increasingly, the RSPB is taking the interests of moths into account when planning management, as at Minsmere and Walberswick, which support rare reed bed moths.

Bird observatories are strategically placed to receive rare migrant moths as well as birds, and most run a light trap nightly to record them. The surrounding area often supports unusual breeding moths too. Portland is home to the Beautiful Gothic *Leucochlaena oditis*, Feathered Brindle *Aporophyla australis* and of course the Portland Ribbon Wave *Idaea degeneraria*. Dungeness has a whole galaxy of rare moths, several of them found nowhere else in Britain, and others very localised. I went there desperately hoping to see the White Spot *Hadena albimacula*, but need not have worried: it was the most numerous species in the trap. Sandwich Bay has the Bright Wave, Gibraltar Point the Marsh Moth and Scarce Pug, Spurn has coastal species found nowhere else in Yorkshire, and so on. For the moth enthusiast, bird observatories are well worth a visit or a longer stay.

Needless to say, every nature reserve set up to preserve and manage interesting habitat, whether run by local authorities, English Nature, Scottish Natural Heritage or county wildlife trusts, will benefit moths even if this was not part of the original intention. When southern heathland is protected for the Smooth Snake and Dartford Warbler, the Shoulder-striped Clover and the Speckled Footman are also helped. If chalk downland is preserved for rare orchids, its characteristic moths are safer too. When the National Trust acquires long lengths of coastline for scenic reasons, this guards the haunts of cliff and sand-dune moths, especially as coastal habitats often need little management. They tend to be self-renewing through erosion, deposition and the stunting effects of salt spray.

Of course, many reserves hold no scarce moths, but this does not detract from their value for this group. Perhaps conservation is most effective when it protects species *before* they become scarce, instead of stepping in only after they have declined so far that the prognosis is hopeless, as arguably happened with the Large Blue butterfly and the Essex Emerald.

The wider countryside

Nature reserves cover only a tiny percentage of the land area of Britain. The rest of the land is not managed primarily for its wildlife. However, attitudes generally are becoming more favourable towards conservation: a concern for the environment is now mainstream and respectable, no longer seen as just the province of cranks. Most gardeners, and some farmers, try to co-exist with wildlife rather than destroy it. Local authorities, the Forestry Commission, water companies, industry and commerce are all keen to demonstrate their Green credentials. Areas under Ministry of Defence control are among our finest wildlife refuges: for example, the best surviving Breckland, and chalk grassland on Salisbury Plain. The requirements of military training – large areas of varied, semi-natural habitat – dovetail perfectly with the needs of wildlife conservation, hence there are 264 Sites of Special Scientific Interest on Ministry of Defence land (Dickinson *et al.*, 1999). In its visitor centre, the grim enormous Dungeness power station now mounts a display of research that Nuclear Electric has funded to safeguard the Sussex Emerald, and we no longer consider that surreal.

Concern for the human environment benefits wildlife too. Smoke pollution, and the killer smogs of my own childhood, have been reduced by the Clean Air Acts and other measures prohibiting the emission of toxic fumes. Persistent pesticides such as DDT have been all but eliminated from the food chain, and other dangerous agricultural chemicals such as dieldrin withdrawn or more strictly controlled, partly through consumer pressure. Exhaust gases from motor vehicles are much cleaner thanks to more efficient engines and additives. Industry is no longer allowed to tip toxic waste and heavy metal contaminants, discharge dioxins into the atmosphere, or pollute rivers and waterways. It is easy to forget how rife and unregulated these practices were only 50 years ago. It used to be said in Salford that no one lived long enough to drown after falling into the River Irwell. Undoubtedly our towns and cities are cleaner nowadays, and the countryside is at least no worse than it was.

Modern farmland, however, is undeniably less friendly for wildlife than at any time in the past. For moths, some of the worst damage was done long ago, during the great agricultural improvements of the late eighteenth and early nineteenth century, especially those involving drainage of extensive wetlands. Later, southern heathland and chalk downland suffered from the abandonment of the agricultural systems that maintained them, disappearing under the plough or being taken over by scrub and eventually woodland. It is slightly ironic when farmers are blamed for destroying wildlife habitats that were created by farming in the first place. The removal of hedges to make bigger fields is deplored, but the hedges themselves date from enclosures of open farmland in the past. The medieval three-field system had no need of them (O'Connor and Shrubb, 1986).

As a result of these and other changes, the productive areas of the average farm are now of little use to moths. Thanks to increasingly sophisticated herbicides and pesticides, arable fields have become barren deserts that support hardly anything apart from the chosen crop. Improved pasture is often little better, and any moths found there tend to be the commonest and most ubiquitous species. It is hard to see how this could be remedied. Whereas altering the timing of agricultural operations can help birds – leaving stubble for winter food, delaying silage cuts so that broods can fledge – moths are not so easily accommodated. Most species require a full year to complete their life cycle. It might do more harm than good if females were tempted to lay on unsprayed weeds, only for these to be ploughed a few months later.

Fortunately, moths are good at surviving in small pockets and corners: along fence lines, hedges and ditches, by the side of railways, roads and tracks, in shelter belts and pheasant copses, anywhere that scraps of suitable habitat still remain. It is sad that even these remnants are still being destroyed. On my Sussex patch, one hillside that had survived because it was too steep for farm machinery was sprayed with herbicide plus fertiliser from the air. A colony of Adonis Blue butterflies and all the downland moths and flowers, which had probably been there for centuries, were destroyed within an afternoon. The farmer, a good and kindly man, did not do this from malice: he was merely improving his grazing.

Plate 128:
Considered a rarity in the nineteenth century, the Kent Black Arches *Meganola albula* has since spread over much of southern England. The raised scale-tufts on the forewings are characteristic of the Nolidae. (Norfolk).

It is hard to be critical of farmers, rather than the system, if they are forced by economic necessity to improve their land to the detriment of its wildlife. There can be less sympathy when the gradual, piecemeal destruction of the small remaining wild areas in the countryside appears to have no obvious commercial motive. I suspect many such 'improvements' are undertaken from a sense of duty, carried over from the days when more food was desperately needed, or for the pride of being seen and admired by neighbours as a neat, tidy, efficient farmer. Often, the cost of the work outweighs any conceivable financial return. A farmer of my acquaintance was determined to cultivate a narrow scrubby bank bordering a large arable field. At best, it represented less than a thousandth of the area of his land. Low scrub that held breeding Lesser Whitethroats and the Kent Black Arches (Plate 128) was bulldozed and burnt. The bank was ploughed and sown with difficulty, but harvesting proved even trickier. The huge combine was forced to transcribe endless figures of eight, cutting only on the downward diagonal stroke. This took 1½ hours, and inevitably the machine needed costly repairs, all for half a hectare of barley. Nor is this an isolated incident. I have seen disproportionate effort and expense go into draining the last boggy corner of a meadow, or canalising a ditch that was already functioning perfectly, because that is what efficient farmers ought to do. Changing such attitudes will be difficult. Even I am torn between good husbandry and nature conservation when managing our own small croft.

Some agricultural changes in recent years have helped moths in minor ways. The ban on straw burning in England is one, not that there are many

Plate 129 Feeding on Flixweed *Descurainia sophia* at the edge of crops, the Grey Carpet *Lithostege griseata* is one of several Breckland moths to have been helped by conservation headland initiatives. (Norfolk).

moths or caterpillars in the stubble itself. At least it has prevented hedge bottoms and field margins being 'accidentally' burnt in the process. Although farmers were supposed to plough a wide firebreak around the edges of fields, often this requirement was ignored, especially if the field was not in view from a public right of way. In any case, the heat could be so intense that trees and hedges downwind of the flames were scorched in spite of the firebreak. This happened in 1991 at the main British colony of the Barberry Carpet, damaging three-quarters of the Barberry bushes that supported this endangered and protected moth (Waring, 2000).

In 1988, European Community grain surpluses resulted in set-aside schemes to take arable land out of production. It was soon realised that this had obvious potential benefits for wildlife, especially birds. In Norfolk, some of the set-aside took the form of strips, 6 metres wide, around the edges of fields. These were cultivated annually, but not sown with crops. Especially in areas with poor soils and a good seed bank, common arable weeds flourished in the fallow ground. Studies by the Game Conservancy showed that breeding success of the Grey Partridge improved. Fortuitously, several of the rare Breckland moths that used the weeds as food plants benefited too, especially the Tawny Wave (Plate 125), Grey Carpet (Plate 129) and Marbled Clover, doubtless providing food for the partridge chicks. Now part of the conservation headlands scheme, and funded by the Ministry of Agriculture, the strips have survived the ending of set-aside (GM Haggett, personal communication).

However, the Breck is perhaps a special case. Few if any scarce moths are associated with arable farmland elsewhere in Britain, except the Brighton Wainscot. Indeed, the outlook for moths is usually better when farmland is lost to house building or new roads, a withering indictment. The extensive cuttings and embankments required by modern dual carriageways provide excellent new breeding areas. Being linear, they act as corridors linking otherwise isolated patches of suitable habitat. Free from sprays and fertilisers, weeds of disturbed ground flourish in the early years: mayweed for the Chamomile Shark *Cucullia chamomillae*, hawk's-beard for the Broad-barred White. If the verge is mown annually at an appropriate time of year, herb-rich grassland can be created, almost the equivalent of a traditional hay meadow.

The value of such verges for plant communities has long been known, and some are managed sympathetically by local authorities as a result, but they are increasingly recognised as being good for moths too. Scarce species associated with road verges include the Striped Lychnis in Buckinghamshire, dependent on the Dark Mullein that grows there (Albertini *et al.*, 1997). Near Lewes in Sussex, the south-facing cuttings of a new dual carriageway through the chalk looked stark and bare when created in the early 1970s, but 15 years later a splendid downland turf had begun to form, with much of its associated insect fauna. There were Bee Orchids and Autumn Lady's-tresses, Glow-worms and crickets, Adonis Blue and Marbled White butterflies. The moths were equally interesting and characteristic. They included the Scarce Forester (Plate 61), a Red Data Book species. Needless to say, all the plants and insects mentioned above had long since disappeared from the improved farmland out of which the new road had been carved.

Conservation and the collector

Collecting is widely perceived as a threat to endangered wildlife, and for some plant or animal groups with a low reproductive capacity this is undoubtedly true. Lepidoptera, however, are able to replenish their populations very quickly. If the habitat is right, they can withstand a high level of predation. In effect, collecting equates to this. There is little evidence, other than anecdotal, that over collecting has ever brought about the extinction of a butterfly or moth. Equally, it is impossible to prove that mortality caused by collectors has never tipped the balance when compounded with loss of habitat.

Nevertheless, most observers instinctively feel that particularly vulnerable moths should not be collected, and support the legal protection offered to a few selected species. A minority would go further, and oppose the taking of any moth except for bona fide scientific purposes. Hence the tendency, in these politically correct times, to describe moths destined for a collection as 'voucher specimens'. Without doubt the latter are sometimes necessary, but the term is stretched when applied to unmistakable species taken from a well-documented site, as not infrequently happens.

Having been a fanatical collector in my own youth, I well understand the attraction. Its hold on me was broken very suddenly. In the middle of a session high on the South Downs, I questioned the value and purpose of my activities and went home, not even bothering to set the moths caught that day. However, I would certainly feel justified in taking up collecting again in a part of the globe where the moth fauna is poorly known. Reference collections are needed in Britain too, both nationally and locally. When visiting friends who are collectors, I always have a mental list of species for examination, to refresh my memory of their identification features. Otherwise, photography serves as my own record of moths found or reared.

Hopefully, whether or not to collect any but the most threatened moths will always remain a personal decision, not one imposed by restrictive legislation. There is much to be said for Morris's (1983) view that Lepidoptera can be 'a self-renewing biological resource' if wisely managed for our interest or pleasure. For those who do wish to form a collection, restraint and responsibility are the modern keynotes. The JCCBI has issued a code of practice in an effort to curtail the undoubted abuses and excesses of the past. The code covers damage to the environment by unacceptable operating methods, as well as the welfare of insect populations themselves. Finally, it should be remembered that specimens without data are of little use: it is a waste to kill and set a moth then omit to label it with the date and exact locality of capture.

Helping to increase public awareness of moths

Plate 130: Dark Bordered Beauty *Epione vespertaria* male. Confined in Britain to a handful of isolated colonies, this pretty moth is clearly vulnerable to adverse habitat changes. (Aberdeenshire).

Image is all-important today. Butterflies have a good one: bright, colourful, sun loving. Moths do not. They are considered to be creatures of the dark, only coming to notice when they make nuisances of themselves by damaging fabrics, or blundering around lighted rooms at night as uninvited guests.

Accordingly, a little public relations work is appropriate whenever the opportunity presents itself. This is not difficult: moths are fascinating, and it is easy to convey this. Anyone who runs a light trap or wanders round with a net peering closely at fences and tree trunks is sure to attract a few curious glances and questions. Simply being open and friendly, and showing some of the more attractive moths found, will suffice. Few people have ever really looked at a moth, and when they do they are often more interested than they had expected to be. The farmer who owns the Aberdeenshire site for the Dark Bordered Beauty (Plate 130) was most reluctant to get out of his truck to examine one sitting on a Bracken frond, being more concerned with searching for a lost cow. He succumbed to Mark Young's blandishments with an

obvious bad grace, but was charmed in spite of himself, declaring it to be a bonny little moth with a very neat pattern, all its markings just so. Maybe that small episode did more for its conservation than any number of Action Plans.

Primary schools usually welcome caterpillars for natural history classes. Obviously, the large, showy ones are best. Pupae are even more exciting, especially if they are due to hatch soon. Emperor Moth, Puss Moth, Elephant Hawk-moth (Plate 131) and Garden Tiger are among those common and widely distributed enough to make suitable teaching aids. It is essential to provide the right equipment plus detailed instructions for their care, and ensure suitable arrangements for weekends and holidays if disasters are to be avoided. Those with suitable slides might consider giving illustrated lectures to school natural history societies, local clubs, or just mildly interested friends.

Often, good sites for moths are destroyed unnecessarily through ignorance, when a simple letter to the authority concerned might have averted this. Local councils often 'tidy up' areas of overgrown land because a few busybodies have complained of its neglected appearance. Often, they are glad of a reason not to do so, saving the trouble and expense, especially if no safety risk is involved. There are times when conserving moths costs only the price of a stamp.

Plate 131: Breathtakingly beautiful species like the Elephant Hawk-moth *Deilephila elpenor* can be used to promote a greater appreciation of moths amongst the general public and in schools. (Banffshire).

The future and climatic change

It is now generally accepted that we live in a period of global warming, although there is dispute about its cause and how much is due to human activities. As the fossil records show, climates have always fluctuated; currently, we live in an interlude between ice ages. If global warming does continue, it will almost certainly enhance Britain's moth fauna. Already there is evidence that this is happening. Almost annually, new colonists arrive from the European continent, and throughout Britain the trend is for distributions to shift gradually northwards. Aberdeenshire, with a long history of recording, has gained many more moths than it has lost in recent decades and is continuing to do so. Among the latest arrivals is the Satin Beauty.

These processes are likely to continue. Any predictions can only be tentative, but perhaps flight times will become earlier as temperature thresholds alter in spring. Species that were previously univoltine may fit in second broods. Migrant moths that cannot quite establish permanent breeding populations in Britain at present, like the Bedstraw Hawk-moth *Hyles gallii* and the Delicate, might manage to do so in East Anglia and Devon respectively. Perhaps former residents like the Flame Brocade will return; it appears that the Small Ranunculus *Aetheria dysodea* has already done so. Unexpected new colonists are likely to include several of that basically southern group Sterrhinae, the waves, and no doubt yet more pugs.

Will there be a price to pay for such gains? Is Britain set to lose some of its more northern, Arctic-Alpine species? There is little evidence of that so far – all the Scottish moorland and mountain specialities (what might be called the 'Ptarmigan moths') seem to be thriving, although hard data are lacking. Some of the effects of climatic warming are likely to be very subtle, operating via their effect on habitat rather than directly. Perhaps the competitive balance between species will be altered, favouring those better able to tolerate, or adapt to, the new conditions. Genetic diversity will be an asset here. For the lepidopterist, these are interesting times.

Appendix 1

Species Totals of Macro-moths, and Micros if known, for the Watsonian Vice-Counties of Great Britain

		Macro-moths	*Micro-moths*	*Source*
1	West Cornwall	516	649	Dr Frank Smith
2	East Cornwall	554	668	Dr Frank Smith
3	South Devon	629	–	RF McCormick
4	North Devon	515	–	RF McCormick
5	South Somerset	610	–	Jim Lidgate
6	North Somerset	569	–	Ray Barnett
7	North Wiltshire	601+	–	estimate: no information received
8	South Wiltshire	601+	–	estimate: no information received
9	Dorset	687	–	Peter Davey
10	Isle of Wight	613	–	Barry Goater
11	South Hampshire	678	–	Barry Goater
12	North Hampshire	626	–	Barry Goater
13	West Sussex	662	–	Colin Pratt
14	East Sussex	671	–	Colin Pratt
15	East Kent	665	–	Eric Philp
16	West Kent	589	–	Eric Philp
17	Surrey	623	1088	Graham A Collins; RM Palmer
18	South Essex	608	917	Brian Goodey
19	North Essex	588	939	Brian Goodey
20	Hertfordshire	551	–	Colin W Plant
21	Middlesex	576	761	Colin W Plant
22	Berkshire	612	–	Martin Harvey
23	Oxfordshire	634	1045	John Campbell
24	Buckinghamshire	620	720	Martin Albertini

		Macro-moths	*Micro-moths*	*Source*
25	East Suffolk	580	–	estimate: Tony Pritchard
26	West Suffolk	560	–	estimate: Tony Pritchard
27	East Norfolk	548+	600+	Ken Saul
28	West Norfolk	550+	515+	Ken Saul
29	Cambridgeshire	552	–	Raymond Revell
30	Bedfordshire	576	–	Len Field
31	Huntingdonshire	591	718	Barry Dickerson
32	Northamptonshire	625	–	John Ward
33	East Gloucestershire	600	880	estimates: Roger Gaunt
34	West Gloucestershire	615	900	estimates: Roger Gaunt
35	Monmouthshire	532	576	Martin Anthoney
36	Herefordshire	528	818	Michael Harper
37	Worcestershire	538	760	Dr ANB Simpson
38	Warwickshire	564	–	David Brown
39	Staffordshire	514	618	David Emley
40	Shropshire	624	639	Adrian Riley
41	Glamorgan	589	667	Barry Stewart & DRW Gilmore
42	Breconshire	464	477	Norman Lowe
43	Radnorshire	354	–	Caroline Moscrop
44	Carmarthenshire	528	423	Steve Lucas
45	Pembrokeshire	501+	–	my estimate: data 'inaccessible'
46	Cardiganshire	556	300+	Adrian Fowles
47	Montgomeryshire	461	–	Andy Law
48	Merionethshire	480	–	Andrew Graham
49	Caernarvonshire	505	–	Debbie Evans
50	Denbighshire	495	500+	Bryan Formstone
51	Flintshire	460	–	Geoff Neal
52	Anglesey	401+	–	my estimate (Doug Murray: 279+)
53	South Lincolnshire	504	614	Rex Johnson
54	North Lincolnshire	593	765	Rex Johnson
55	Leicestershire	547	620	Adrian Russell
56	Nottinghamshire	564	566	Sheila Wright
57	Derbyshire	490	644	Ian Viles
58	Cheshire	544	–	Ian Rutherford
59	South Lancashire	490	–	Stephen M Palmer
60	West Lancashire	510	–	Stephen M Palmer
61	South-east Yorkshire	499	–	Barry Spence
62	North-east Yorkshire	462	–	Barry Spence
63	South-west Yorkshire	482	–	Barry Spence

		Macro-moths	Micro-moths	Source
64	Mid-west Yorkshire	479	–	Barry Spence
65	North-west Yorkshire	329	–	Barry Spence
66	Durham	453	570	Alan & Jeri Coates
67	South Northumberland	398	–	Nicholas Cook
68	North Northumberland	306	–	Nicholas Cook
69	Westmorland	513	–	Neville Birkett
70	Cumberland	537	–	Neville Birkett
71	Isle of Man	420	420	Gordon Craine
72	Dumfriesshire	418	–	Peter Norman
73	Kirkcudbrightshire	423	–	Peter Norman
74	Wigtownshire	368	–	Peter Norman
75	Ayrshire	330	–	estimate*
76	Renfrewshire	201+	–	estimate*
77	Lanarkshire	230+	–	estimate*
78	Peeblesshire	220+	–	estimate*
79	Selkirkshire	230+	–	estimate*
80	Roxburghshire	390	–	estimate*
81	Berwickshire	450	–	estimate*
82	East Lothian	330	–	estimate*
83	Midlothian	330	–	estimate*
84	West Lothian	330	–	estimate*
85	Fifeshire	360	–	estimate*
86	Stirlingshire	401+	–	estimate*
87	West Perthshire	375	–	estimate*
88	Mid Perthshire	360	–	estimate*
89	East Perthshire	390	–	estimate*
90	Angus	301+	–	estimate*
91	Kincardineshire	309	323	RM Palmer
92	South Aberdeenshire	341	479	RM Palmer
93	North Aberdeenshire	294	347	RM Palmer
94	Banffshire	306	262	Roy Leverton
95	Moray	356	–	David Barbour
96	East Inverness-shire	327	–	Steven Moran
97	West Inverness-shire	201	–	Steven Moran
98	Argyll Main	360	–	estimate*
99	Dunbartonshire	320	–	estimate*
100	Clyde Isles	275	–	estimate*
101	Kintyre	230	–	estimate*
102	South Ebudes	215	–	estimate*

		Macro-moths	Micro-moths	Source
103	Mid Ebudes	315	–	estimate*
104	North Ebudes	275	–	Steven Moran
105	West Ross	275	–	Steven Moran
106	East Ross	301	–	Steven Moran
107	East Sutherland	175	–	Steven Moran
108	West Sutherland	169	–	Steven Moran
109	Caithness	248	–	Rosie (1976) & Adrian Riley (Rothamsted)
110	Outer Hebrides	101	–	(Lewis only) John Trevor
111	Orkney Islands	213	**188**	Sidney Gauld
112	Shetland Islands	146	**133**	Mike Pennington

Key

Totals in bold are actual counts, those not in bold are estimates.

*For Scottish vice-counties 75–90 and 98–103, data were extracted from the dot maps in the various volumes of *Moths and Butterflies of Great Britain and Ireland.* To allow for under-recording, and lack of maps for the Geometridae, the total so obtained was multiplied by a correction factor, derived from a similar exercise for vice-counties 91–94 where the true totals were known. These estimates must be treated with caution.

METHODS

All county recorders listed in Waring (1999), or their replacements where relevant, were asked for the total number of species of macro-moths reported at least once from their vice-county since 1901. Macro-moths were defined as all those species listed in Skinner (1998). It was left to the recorder to decide the validity of records. Separate totals were requested for micros, if available. The willing co-operation of the recorders listed is much appreciated.

Where no total for macro-moths was available, this was estimated using neighbouring counties as a guide. Allowance should be made for the varying degree of coverage in the less well-populated vice-counties. The micros in particular are very under-worked. For macro-moths, the general picture seems reasonably accurate.

Appendix 2

Presumed Gains and Losses of Resident British Microlepidoptera since records began

APPARENT EXTINCTIONS

Species	Former distribution	Last sighting, if known
Ectodemia beirnei	S England	1935
Stigmella desperatella	locally in England	no recent record
S. torminalis	Herefordshire	late 19th Century?
Opostega spatulella	Essex and Somerset	19th Century, but perhaps over-looked
Tischeria gaunacella	Essex	19th Century
Tenaga nigripunctella	England and Wales	early 20th Century
Cephimallota angusticostella	SE England	1899
Paraleucoptera sinuella	Scottish Highlands	1950, but probably overlooked
Lyonetia prunifoliella	widespread in England	1900
Heliodines roesella	S & E England	1820s
Kessleria fasciapennella	SE Scotland	mid-19th Century
Ypsolopha asperella	Herefordshire & Dorset	1860s
Augasma aeratella	S & E England	1956
Coleophora vulnerariae	E Kent	late 1880s (transitory resident?)
C. antennariella	Oxfordshire	1927, but possibly overlooked
Dafa formosella	SE England	*c* 1840
Borkhauseria minutella	widespread in Britain	1928
Depressaria depressana	S & SE England	19th Century (transitory resident?)
Aristotelia subdecurtella	East Anglia	no recent record
Bryotropha figulella	E Suffolk	no recent record
Gnorimoschema streliciella	NW England, Scotland	1907

Species	Former distribution	Last sighting, if known
Sophronia humerella	locally in England	old unconfirmed records
Mesophleps silacella	SE England	no recent record
Cosmopterix schmidiella	S England	no recent record
Euclemensia woodiella	S Lancashire	1829
Archips betulana	East Anglia	1900 (transitory resident?)
Choristoneura lafauryana	W Norfolk	1962 (transitory resident?)
Olethreutes doubledayana	E England	no recent record
Pristerognatha penthinana	NW England	1914
Gypsonoma nitidulana	Scottish Highlands	1908, but probably overlooked

Notes. This list is based on information extracted from Bradley (1998) and Emmet (1992), plus valued help from Mark Young and DJL Agassiz. However, since any such list is subjective and debatable, the final selection of species was my own. A few of the above moths may have been temporary colonists rather than established residents. Secondly, it is difficult to prove a negative: microlepidoptera are easily overlooked, and it is possible that some of these species may be refound.

APPARENT GAINS

Species	First sighting	Distribution	Food plant	Origin
Etainia decentella	1931	England & Wales	Sycamore	Europe
Ectoedemia erythrogenella	1973	coastal SE England	bramble	S Europe
E. amani	1994	SE England	?	?
Bohemannia auriciliella	1973	SE England	birch	Europe
Stigmella speciosa	1957	England and Wales	Sycamore	C Europe
S. suberivora	1928	S England	Quercus ilex	S Europe
Emmetia heinemanni	1984	SE England	bramble, agrimony	?
Lampronia flavimitrella	1974	S England	bramble	?
Psychoides filicivora	1909	England, Wales, Ireland	ferns	Far East
Oinophila v-flava	?	Scilly Isles	detritus, fungus	?
Caloptilia rufipennella	1970	Britain & Ireland	Sycamore	Europe
C. azaleella	?	S England, S Scotland	azaleas	E Asia
Phyllonorycter platani	1990	SE England, Midlands	Plane	C Europe
P. leucographella	1989	S England, Midlands	Pyracantha	S Europe
P. strigulatella	1928	England	Grey Alder	C or N Europe
P. sagitella	1955	Midlands, Wales	Aspen	Europe

Species	First sighting	Distribution	Food plant	Origin
Phyllocnistis xenia	1974	S England	Grey Poplar	Europe
Tinagma balteolella	1975	SE England	*Echium*	Europe
Argyresthia trifasciata	1982	SE England, Scotland	conifers	Europe
A. cupresella	1997	SE England	conifers	N America
Coleophora fuscicornis	1973	SE England	*Vicia*	S Europe
C. linosyridella	1978	SE England	Sea Aster	Europe
C. aestuariella	1981	SE England	Annual Sea-blite	S Europe
Elachista littoricola	1982	S England	*Elytrigia*	Europe
Bisigna procerella	1976	SE England	lichens	?
Parocystola acroxantha	1908	local, England	decaying leaves	Australia
Ethmia terminella	?	SE England	*Echium*	S Europe
Argolamprotes micella	1963	SW England, S Wales	*Rubus*	C Europe
Monochroa hornigi	1961	S England	*Persicaria*	Europe
M. niphognapha	1984	SE England	*Persicaria*	Europe
Athrips rancidella	1971	SE England	*Cotoneaster*	N America
Coleotechnites piceaella	1952	S England	pine	N America
Teleiodes alburnella	1927	Britain	birch	C Europe
Gelechia senticetella	1988	S England	Juniper	Europe
G. sabinella	1971	S England	Juniper	Europe
Blastobasis lignea	1902	Britain & Ireland	detritus	Madeira
B. decolorella	1946	Britain	detritus	Madeira
Cochylis molliculana	1993	S England	*Picris*	Europe
Cacoecimorpha pronubana	1905	Britain & Ireland	garden shrubs	S Europe
Ptycholomoides aeriferanus	1951	England & Wales	Larch	C Europe
Epiphyas postvittana	1936	England & Wales	garden shrubs	Australia
Adoxophyles orana	1950	S & C England	fruit trees	Australia?
Lozotaeniodes formosanus	1945	S England & Wales	pine	Europe
Acleris abietana	1965	N England, Scotland	conifers	N & C Europe
Crocidosema plebejana	1900	S Britain & Ireland	Malvaceae	S Europe
Epiblema grandaevana	1890?	NE-SE England	Colt's-foot	Baltic
Eucosma metzneriana	1977	S England	*Artemisia*	Europe
Pammene aurita	1943	England & Wales	Sycamore	N & C Europe
Cydia medicaginis	1970	S England	*Medicago*	Europe
C. pactolana	1965	S England	conifers	Europe
C. illutana	1975	S England	Larch	Europe
C. conicolana	1934	England & Wales	pine	Europe
Evergestis limbata	1994	S England	*Alliaria*	S Europe
Ostrinia nubilalis	1930s	England, Wales, Ireland	Mugwort	S Europe

Species	First sighting	Distribution	Food plant	Origin
Phlyctaenia perlucidalis	1951	England & Wales	thistles	Europe
Dioryctria schuetzeella	1980	SE & S England	*Picea*	C Europe
Ancylosis oblitella	1950	S & C England	*Chenopodium*	Europe/Asia
Stenoptilia millieridactyla	1937	Britain & Ireland	saxifrages	C Europe

Notes. The above information has been extracted from the various volumes of *Moths and Butterflies of Great Britain and Ireland*, Bradley *et al.* (1973 & 1979) and Bradley (1998). Mark Young provided helpful advice and suggestions, but again the final selection of species was my own. Adventives without breeding populations in the wild have been excluded. Sometimes there is doubt as to whether the species was an overlooked resident or a newly arrived colonist, and if so, whether it has successfully established itself. In these cases I have gone by the balance of probabilities.

Appendix 3

Lepidoptera Resident in Britain believed to have been Introduced to North America

DEFINITION

The species must be established in the wild, thus some cosmopolitan pests of stored products, and others only inhabiting heated buildings, are excluded. There must be strong evidence for artificial introduction. It is not suggested that Britain, as opposed to Europe or even the Old World in general, was necessarily the source for all of the introductions. Indeed, some introductions to California may have arrived via the Far East. The systematic arrangement and numbering follows Bradley (1998).

136 *Lampronia rubiella* New Brunswick, Canada by 1936, since spread rapidly. *MBGBI*.

161 *Zeuzera pyrina* by 1879, now occasional pest in eastern USA, Covell, *MBGBI*.

186 *Psyche casta* now along New England coast, Covell, *MBGBI*.

215 *Nemapogon grannella* European Grain Moth, throughout North America in fields as well as indoors, Covell.

236 *Tineola bisselliella* Common Clothes Moth, an ancient introduction, Powell.

253 *Ochsenheimeria vacculella* Cereal Stem Moth, introduced before 1964, now pest of cereals, Covell, *MBGBI*.

277 *Oinophila v-flava* California, perhaps since 1880s, now common, stored food, detritus, Powell.

285 *Caloptilia azaleella MBGBI*.

326 *Phyllonorycter blandcardella* introduced to apple-growing regions, *MBGBI*.

370 *Sesia apiformis MBGBI*.

373 *Synanthedon tipuliformis* introduced with currant plants, *MBGBI*.

389 *Choreutis pariana* Apple Leaf Skeletonizer, New England by 1917, now everywhere, Covell.

418 *Argyresthia conjugella* introduced in 1897, Covell.

420 *A. pruniella* USA, *MBGBI*.

425 *Yponomeuta padella* North America, *MBGBI*.

437 *Swammerdamia caesiella* Watson and Whalley.

438 *S. pyrella* Watson and Whalley.

453 *Ypsolopha dentella* Watson and Whalley.

464 *Plutella xylostella* introduced by 1850, now minor pest of cabbages in eastern USA, Watson and Whalley.

465 *P. porrectella* eastern USA, more recently (1961) California, Powell.

485 *Schreckensteinia festaliella* distribution and food plants suggest introduction.

517 *Coleophora frischella* imported in clover seeds, 1966, now a minor pest, Watson and Whalley.

526 *C. laricella* introduced about 1917, *MBGBI*.

640 *Batia lunaris* West Coast USA, since 1956, under dead bark, Powell.

644 *Borkhausenia fuscescens* California since 1936 (new to N America), dead leaves, birds' nests, Powell.

648 *Endrosis sarcitrella* White-shouldered House Moth, detritus, Powell.

649 *Esperia sulphurella* California since 1966, some spread, fungus, dead wood, Powell.

695 *Agonopteryx alstromeriana* New York 1973, western USA in 1983, very rapid spread since, on *Conium*, Powell.

706 *A. nervosa* California, on broom, Powell.

720 *Ethmia bipunctella* New York 1964, now spreading, on *Echium vulgare* an introduced weed, Powell.

724 *Metzneria lappella* eastern North America, Covell.

760 *Exoteleia dodecella* Ontario, Covell.

857 *Anarsia lineatella* Peach Bud Moth, serious pest in southern States, Covell.

870 *Oegoconia quadripuncta* California, leaf litter, Powell.

872 *Symmoca signatella* adventive in UK but widespread in California since 1943, dry vegetable matter, Powell.

903 *Glyphipteryx linneella* New Jersey, Watson and Whalley.

947 *Aethes smeathmanniana* widespread, Watson and Whalley.

970 *Pandemis cerasana* Vancouver, fruit trees and shrubs, Powell.

972 *P. heparana* Vancouver, fruit trees and shrubs, Powell.

977 *Archips podana* orchards, Bradley *et al.*

985 *Cacoecimorpha pronubana* introduced to UK and to USA, Portland, Oregon since 1964, Powell.

994 *Clepsis consimilana* Long Island since 1939, now Connecticut, Powell.

1010 *Ditula angustiorana* Powell.

1015 *Eulia ministrana* distribution, Europe and east coast of America, suggests introduction.

1016 *Cnephasia longana* Omnivorous Leaf-tier, introduced 1927, now serious pest of strawberries, Bradley *et al.*

1036 *Croesia forsskaleana* Long Island 1932, now spread into eastern USA, on *Acer*, Powell.

1037 *Acleris holmiana* Vancouver, Rosaceae, Powell.

1039 *A. comariana* imported in azaleas, 1924, Bradley *et al.*

1048 *A. variegana* Powell.

1177 *Epiblema rosaecolana* eastern USA, Bradley *et al.*

1210 *Rhyacionia buoliana* European Pine-shoot Moth, Bradley *et al.*

1257 *Cydia nigricana* Bradley *et al.*, Watson and Whalley, *MBGBI*.

1261 *C. pomonella* Bradley *et al.*, Covell.

1288 *Alucita hexadactyla* Covell, Watson and Whalley.

1375 *Ostrinia nubilalis* European Corn-borer, introduced *c.* 1908, now probably the most serious corn pest, Covell.

1426 *Achroia grisella* Lesser Wax Moth 1890s, eastern USA, Powell.

1479 *Plodia interpunctella* Indian Meal Moth, only adventive in UK, long introduced to North America, Powell.

1527 *Thymelicus lineola* Canada, 1910, range still spreading, a pest of hay, Watson and Whalley, *MBGBI*.

1550 *Pieris rapae* introduced in nineteenth century, now a crop pest across North America, Watson and Whalley.

1799 *Operophtera brumata* Nova Scotia and New Brunswick, believed introduced, Watson and Whalley.

1860 *Chloroclystis rectangulata* Nova Scotia, 1972, Watson and Whalley.

1927 *Hyloicus pinastri* frequently recorded as an accidental introduction, Watson and Whalley.

1984 *Macroglossum stellatarum* Recorded in N America, but status uncertain, per Honey via Internet.

1986 *Hyles euphorbiae* introduced to control *Euphorbia* weeds in western Canada, Watson and Whalley, *MBGBI*.

2029 *Euproctis chrysorrhoea* since 1897, once major pest, now confined to coastal islands, Covell, *MBGBI*.

2030 *E. similis* distribution suggests this may also be an importation.

2031 *Leucoma salicis* 1920, now sometimes a pest northern USA and Canada, Watson and Whalley, *MBGBI*.

2034 *Lymantria dispar* escaped in 1869, serious pest, range still expanding, Covell, Watson and Whalley, *MBGBI*.

2047 *Eilema complana* introduced by 1903, redescribed as *Crambidia alleghenien-sis*! Watson and Whalley.

2069 *Tyria jacobaeae* released in N America to control ragwort, *MBGBI*.

2107 *Noctua pronuba* recent introduction, already locally abundant, per Internet

2109 *N. comes* per Internet.

2176 *Cerapteryx graminis* 1917.

2223 *Calophasia lunula* per Internet.

2331 *Apamea unanimis* per Internet.

2336 *A. ophiogramma* per Internet.

2361 *Hydraecia micacea* Potato Stem Borer, a minor pest, Covell.

2387 *Caradrina morpheus* per Internet.

Notes. This list is almost certainly very incomplete. There are surprisingly few leaf-miners, for instance. In general, only those introduced species that have brought attention upon themselves by becoming pests have been documented in the literature. Even then, many of the species in the above list are described as introductions by only one or two authors. Although numerous moths have a Holarctic distribution, there are instances where otherwise Western Palearctic moths are found only on the eastern coast of North America. Artificial introduction seems the most likely explanation here. For example, the distributions of many tortricids which are associated with fruit trees and bushes such as *Ribes, Malus, Pyrgus* and *Prunus*, or with ornamental shrubs, follow this pattern. Unless the authors quoted have described these species as introduced I have included only the most obvious candidates. Therefore, it should be possible to extend this list considerably. Also, it might well contain errors, and should be regarded only as a framework for further work.

North American Lepidoptera now established in Britain and the European Mainland

	Parectopa robiniella now in eight mainly southern European countries, Karsholt and Razowski.

Parectopa robiniella now in eight mainly southern European countries, Karsholt and Razowski.

Phyllonorhycter robiniella now in seven mainly southern European countries, Karsholt and Razowski.

409b *Argyresthia cupressella* SE England, 1997, Agassiz and Tuck.

761a *Athrips rancidella* SE England, 1971, on *Cotoneaster*.

759   *Coleotechnites piceaella* S England, on pine.

Hyphantria cunea Fall Webworm, introduced to Hungary, Watson and Whalley; now in 18 central and southern European countries, Karsholt and Razowski.

2203 *Mythimna unipuncta* common in southern Europe but only a scarce migrant to British Isles.

Notes. Two butterflies, the Monarch and the American Painted Lady *Vanessa virginiensis*, are established in the Azores, Canaries and Madeira, but not on the European mainland, where they occur only as scarce migrants.

There may well be other species that qualify for inclusion, but little information is available. I have excluded two china-mark moths from the USA, *Synclita obliteralis* and *Paraponyx obscuralis*, which are among a dozen or so exotics of this sub-family recorded breeding in aquatic greenhouses at Enfield near London, as they have never been found in the wild. Note that many other species of American origin are on the British list solely as adventives, usually transported in vegetable produce. In a few cases, they were found here before being known to science in their native land, so that the type locality is, ironically, in Britain. This should not be allowed to cause confusion.

REFERENCES

Agassiz DJL and Tuck KR (1999) The Cypress Tip Moth *Argyresthia cupressella* Walsingham 1890 (Lepidoptera: Yponomeutidae) new to Britain. *Entomologist's Gazette* 50: 11–16.

Bradley JD (1998) *Checklist of Lepidoptera recorded from the British Isles*. Bradley, Fordingbridge.

Bradley JD, Tremewan WG and Smith A (1973 & 1979) *British Tortricoid Moths* (two vols) Ray Society, London.

Covell CV (1984) *Eastern Moths*. Peterson Field Guides. Houghton Mifflin, Boston and New York.

Karsholt O and Razowski J (eds) (1996) *The Lepidoptera of Europe*. Apollo Books, Strenstrup.

MBGBI, various authors (1979–96) *The Moths and Butterflies of Great Britain and Ireland* vols 1, 2, 3, 7(1), 7(2), 9 & 10. Harley, Colchester.

Powell JA (personal communication) much information from numerous sources, for which I am very grateful.

Watson A and Whalley P (1983) *The Dictionary of Butterflies and Moths*. Peerage Books, London.

Appendix 4

British Macro-moths that are frequently active in daylight

Gold Swift *Hepialius hecta*	late afternoon/evening (especially in Scotland?)
Forester Moths (Procridinae)	three species, all diurnal, in sunshine
Burnet Moths (Zygaeninae)	seven species, all diurnal, in sunshine
Festoon *Apoda limacodes*	in hot sunshine (Heath and Emmet, 1985)
Triangle *Heterogenea asella*	afternoon sunshine (Heath and Emmet, 1985)
Clearwings (Sesiidae)	15 species, all diurnal, mainly in sunshine, not often seen
Oak Eggar *Lasiocampa quercus*	males diurnal, midday and afternoon
Fox Moth *Macrothylacia rubi*	males diurnal, late afternoon to evening
Emperor *Saturnia pavonia*	males diurnal, mid-afternoon
Kentish Glory *Endromis versicolora*	males diurnal, mid-morning to early afternoon
Oak Hook-tip *Drepana binaria*	occasionally in afternoon sunshine (males?) around oaks
Barred Hook-tip *D. cultraria*	males frequently fly in sunshine around Beech
Orange Underwing *Archiearis parthenias*	males in particular diurnal in sunshine around birch
Light Orange Underwing *A. notha*	males in particular diurnal in sunshine around Aspen
Purple-bordered Gold *Idaea muricata*	especially at sunrise (South, 1909)
Balsam Carpet *Xanthorhoe biriviata*	late afternoon and evening (Skinner, 1998)
Small Argent and Sable *Epirrhoe tristata*	frequently flies in afternoon sunshine (Skinner, 1998)
Blue-bordered Carpet *Plemyria rubiginata*	late afternoon and early evening
Slender-striped Rufous *Coenocalpe lapidata*	females fly on dry afternoons (Skinner, 1998)
Argent and Sable *Rheumaptera hastata*	diurnal, flies in sunshine
Heath Rivulet *Perizoma minorata*	flies in afternoon sunshine

Pretty Pinion *P. blandiata*	flies in late afternoon
Grass Rivulet *P. albulata*	flies in late afternoon and early evening
Twin-spot Carpet *P. didymata*	mainly diurnal and crepuscular
Haworth's Pug *Eupithecia haworthiata*	occasionally in sunshine around Traveller's Joy
Marsh Pug *E. pygmaeata*	diurnal, in sunshine
Satyr Pug *E. satyrata*	mainly diurnal, especially subspecies *callunaria*
Double-striped Pug *Gymnoscelis rufifasciata*	flies in afternoon sunshine (especially in Scotland?)
Manchester Treble-bar *Carsia sororiata*	often flies in the daytime
Chimney Sweeper *Odezia atrata*	diurnal
Small White Wave *Asthena albulata*	evening
Drab Looper *Minoa murinata*	diurnal
Netted Mountain Moth *Macaria carbonaria*	diurnal
Latticed Heath *Chiasmia clathrata*	flies in sunshine
Rannoch Looper *Itame brunneata*	flies in sunshine
Little Thorn *Cepphis advenaria*	diurnal
Dark Bordered Beauty *Epione vespertaria*	after sunrise, also in evening
Speckled Yellow *Pseudopanthera macularia*	diurnal, in sunshine
Common Heath *Ematurga atomaria*	diurnal
Bordered White *Bupalus piniaria*	males partly diurnal (especially in Scotland?)
Black Mountain Moth *Glacies coracina*	diurnal, in sunshine
Black-veined Moth *Siona lineata*	flies in hot sunshine (Skinner, 1998)
Yellow Belle *Semiaspilates ochrearia*	males often fly in sunshine
Narrow-bordered Bee Hawk *Hemaris tityus*	diurnal, in sunshine
Broad-bordered Bee Hawk *H. fuciformis*	diurnal, in sunshine
Humming-bird Hawk *Macroglossum stellatarum*	diurnal
Scarce Vapourer *Orgyia recens*	male diurnal (female flightless)
Vapourer *O. antiqua*	male diurnal (female flightless)
Dew Moth *Setina irrorella*	male flies at dawn and in afternoon
Red-necked Footman *Atolmis rubricollis*	flies round trees in hot sunshine (Skinner, 1998)
Wood Tiger *Parasemia plantaginis*	diurnal, in sunshine, especially the male
Clouded Buff *Diacrisia sannio*	male flies in hot weather
Muslin *Diaphora mendica*	unusually, female is diurnal but male nocturnal
Ruby Tiger *Phragmatobia fuliginosa*	first brood largely diurnal, as is single brood in north
Jersey Tiger *Euplagia quadripunctaria*	flies in hot sunshine
Scarlet Tiger *Callimorpha dominula*	diurnal
Northern Rustic *Standfussiana lucernea*	mainly nocturnal, but flies in hot sunshine (Skinner, 1998)
Least Yellow Underwing *Noctua interjecta*	flies wildly in late afternoon (males?)

True-lover's Knot *Lycophotia porphyrea*	mainly nocturnal, but often flies in sunshine
Beautiful Yellow Underwing *Anarta myrtilli*	diurnal
Small Dark Yellow Underwing *A. cordigera*	diurnal
Broad-bordered White Underwing *A. melanopa*	diurnal
The Silurian *Eryopigodes imbecilla*	flies in hot afternoon sunshine (Skinner, 1998)
Antler Moth *Cerapteryx graminis*	partially diurnal
Middle-barred Minor *Oligia fasciuncula*	male flies in late afternoon
Cloaked Minor *Mesoligea furuncula*	male flies in late afternoon and early evening
Least Minor *Photedes captiuncula*	male flies in sunshine during afternoon (Skinner, 1998)
Small Wainscot *Chortodes pygmina*	male flies strongly in afternoon (especially in Scotland?)
Large Ear *Amphipoea lucens*	occasionally active by day
Saltern Ear *A. fucosa*	occasionally active by day
Crinan Ear *A. crinanensis*	frequently active by day
The Ear *A. oculea*	frequently active by day
Haworth's Minor *Celaena haworthii*	male flies in afternoon
Small Rufous *Coenobia rufa*	flies in early evening
Small Yellow Underwing *Panemeria tenebrata*	diurnal
Marbled Clover *Heliothis viriplaca*	diurnal
Shoulder-striped Clover *H. maritima*	diurnal (Skinner, 1998)
Silver Y *Autographa gamma*	often flies by day, especially newly arrived migrants
Scarce Silver Y *Syngrapha interrogationis*	often flies by day
Mother Shipton *Callistege mi*	diurnal
Burnet Companion *Euclidia glyphica*	diurnal
Four-spotted *Tyta luctuosa*	mainly diurnal
Small Purple-barred *Phytometra viridaria*	diurnal

Notes. This list is confined to species that are regularly active in full daylight, in one sex at least, of their own accord. Those not described as diurnal may also fly at dusk or at night as well. Numerous other moths are occasionally active by day, for example several of the Plusiinae, but this cannot be described as their usual behaviour. Almost certainly there are regional differences: for instance, the Ruby Tiger is mainly diurnal in NE Scotland. Scarce migrants have been excluded.

Appendix 5

Scientific Names of Species other than Lepidoptera mentioned in the text

PLANTS

Alder *Alnus glutinosa*
Apple *Malus*
Archangel, Yellow *Lamiastrum galeobdolon*
Ash *Fraxinus excelsior*
Aspen *Populus tremula*
Azalea, Trailing *Loiseleuria procumbens*
Barberry *Berberis vulgaris*
Bearberry *Arctostaphylos uva-ursi*
Bedstraw *Galium*
Beech *Fagus sylvatica*
Bilberry *Vaccinium myrtillus*
Bilberry, Bog *Vaccinium uliginosum*
Bindweed, Field *Convolvulus arvensis*
Birch *Betula*
Birch, Silver *Betula pendula*
Bluebell *Endymion non-scriptus*
Bog-myrtle *Myrica gale*
Bracken *Pteridium aquilinum*
Bramble *Rubus fruticosus*
Broom *Cytisus scoparius*
Buckthorn *Rhamnus catharticus*
Buddleia *Buddleia davidii*
Bugle *Ajuga reptans*
Bulrush *Typha*
Burnet, Salad *Sanguisorba minor*
Butcher's-broom *Ruscus aculeatus*
Buttercup *Ranunculus*

Campion, Bladder *Silene vulgaris*
Campion, Red *Silene dioica*
Campion, Sea *Silene maritima*
Campion, White *Silene alba*
Canary-grass, Reed *Phalaris*
Catchfly, Night-flowering *Silene noctiflora*
Catchfly, Nottingham *Silene nutans*
Cherry *Prunus*
Cherry, Bird *Prunus padus*
Chickweed *Stellaria*
Clover *Trifolium*
Cotoneaster *Cotoneaster*
Cowberry *Vaccinium vitis-idaea*
Crowberry *Empetrum nigrum*
Cuckooflower *Cardamine pratensis*
Currant *Ribes*
Currant, Black *Ribes nigrum*
Cypress, Leyland *Cupressocyparis leylandii*
Cypress, Monterey *Cupressus macrocarpa*
Cypress, Lawson's *Chamaecyparis lawsoniana*
Dandelion *Taraxacum*
Delphinium *Delphinium*
Dock *Rumex*
Elder *Sambucus nigra*
Elm *Ulmus*
Elm, Wych *Ulmus glabra*
Fir, Douglas *Pseudotsuga menziesii*

Firethorn *Pyracantha coccinea*
Flag, Yellow *Iris pseudacorus*
Foxglove *Digitalis purpurea*
Goldenrod *Solidago virgaurea*
Gooseberry *Ribes uva-crispa*
Gorse *Ulex*
Grass, Cock's-foot *Dactylis glomerata*
Grass, Marram *Ammophila arenaria*
Hawk's-beard *Crepis*
Hawkweed *Hieracium*
Hawthorn *Crataegus monogyna*
Heather *Calluna, Erica*
Heather, Bell *Erica cinerea*
Hogweed *Heracleum sphondylium*
Honeysuckle *Lonicera periclymenum*
Horse-chestnut *Aesculus hippocastanum*
Hound's-tongue *Cynoglossum officinale*
Ivy *Hedera helix*
Jasmine, White *Jasminum officinale*
Juniper *Juniperus communis*
Knapweed *Centaurea*
Knotgrass *Polygonum aviculare*
Lady's-tresses, Autumn *Spiranthes spiralis*
Larch *Larix decidua*
Larkspur *Consolida ambigua*
Lime *Tilia*
Ling *Calluna vulgaris*
Lyme-grass *Leymus arenarius*
Marjoram *Origanum vulgare*
Mayweed *Matricaria, Tripleurospermum*
Meadow-rue *Thalictrum*
Monk's-hood *Aconitum napellus*
Mugwort *Artemisia vulgaris*
Mullein *Verbascum*
Mullein, Dark *Verbascum nigrum*
Nettle, Stinging *Urtica dioica*
Oak *Quercus*
Orchid, Bee *Ophrys apifera*
Orchid, Marsh *Dactylorhiza*
Orchid, Spotted *Dactylorhiza*
Pear *Pyrus*
Pignut *Conopodium majus*
Pine *Pinus*

Pine, Lodgepole *Pinus contorta*
Pine, Scots *Pinus sylvestris*
Plantain *Plantago*
Plum *Prunus*
Poplar *Populus*
Primrose *Primula vulgaris*
Primrose, Evening *Oenothera*
Privet *Ligustrum*
Ragged-Robin *Lychnis flos-cuculi*
Ragwort *Senecio*
Ragwort, Common *Senecio jacobaea*
Raspberry *Rubus idaeus*
Rattle, Yellow *Rhinanthus minor*
Reed, Common *Phragmites australis*
Rocket, Sweet *Hesperis matronalis*
Rose *Rosa*
Sage, Wood *Teucrium scorodonia*
Sallow *Salix*
Scabious, Devil's-bit *Succisa pratensis*
Snapdragon *Antirrhinum majus*
Sorrel *Rumex*
Sow-thistle *Sonchus*
Spindle *Euonymus europaeus*
Spruce *Picea*
Spruce, Sitka *Picea sitchensis*
Stock, Night-scented *Matthiola bicornis*
Sweet William *Dianthus barbatus*
Sycamore *Acer pseudoplatanus*
Thyme *Thymus*
Toadflax, Common *Linaria vulgaris*
Toadflax, Purple *Linaria purpurea*
Tobacco *Nicotiana*
Tor-grass *Brachypodium pinnatum*
Touch-me-not *Impatiens noli-tangere*
Traveller's-joy *Clematis vitalba*
Trefoil, Bird's-foot *Lotus corniculatus*
Vetch *Vicia, Lathyrus, Hippocrepis*
Violet *Viola*
Willow, Goat *Salix caprea*
Willowherb, Rosebay *Epilobium angustifolium*
Wormwood, Sea *Artemisia maritima*
Yew *Taxus baccata*
Yucca *Yuca*

BIRDS

Auk, Little *Alle alle*
Bee-eater *Merops*
Blackbird *Turdus merula*
Blackcap *Sylvia atricapilla*
Bulbuls: Pycnonotidae
Chaffinch *Fringilla coelebs*
Crake, Spotted *Porzana porzana*
Cuckoo *Cuculus canoris*
Curlew, Stone *Burhinus oedicnemus*
Drongos: Dicruridae
Duck, Long-tailed *Clangula hyemalis*
Dunnock *Prunella modularis*
Fieldfare *Turdus pilaris*
Gannet *Morus bassanus*
Goose, Greylag *Anser anser*
Greenfinch *Carduelis chloris*
Gull, Common *Larus canus*
Hobby *Falco subbuteo*
Lark, Hoopoe *Alaemon alaudipes*
Lark, Wood *Lullula arborea*
Merlin *Falco columbarius*
Nightingale *Luscinia megarhynchos*
Nightjar *Caprimulgus europaeus*
Osprey *Pandion haliaetus*
Owl, Barn *Tyto alba*
Owl, Little *Athene noctua*
Partridge, Grey *Perdix perdix*
Pheasant *Phasianus colchicus*

Pipit, Meadow *Anthus pratensis*
Ptarmigan *Lagopus mutus*
Quail *Coturnix coturnix*
Red Kite *Milvus milvus*
Redpoll *Carduelis flammea*
Redwing *Turdus iliacus*
Robin *Erithacus rubecula*
Rook *Corvus frugilegus*
Sea Eagle *Haliaeetus albicilla*
Siskin *Carduelis spinus*
Skua, Great *Catharacta skua*
Sparrow, House *Passer domesticus*
Sparrowhawk *Accipiter nisus*
Starling *Sturnus vulgaris*
Stonechat *Saxicola torquata*
Swallow *Hirundo rustica*
Thrush, Song *Turdus philomelos*
Tit, Blue *Parus caeruleus*
Tit, Coal *Parus ater*
Tit, Crested *Parus cristatus*
Tit, Great *Parus major*
Tit, Long-tailed *Aegithalos caudatus*
Treecreeper *Certhia familiaris*
Warbler, Dartford *Sylvia undata*
Whitethroat *Sylvia communis*
Whitethroat, Lesser *Sylvia curruca*
Woodpecker, Great Spotted *Dendrocopos major*
Wren *Troglodytes troglodytes*

AMPHIBIANS, REPTILES and MAMMALS

Deer, Red *Cervus elaphus*
Deer, Roe *Capreolus capreolus*
Fox *Vulpes vulpes*
Frog, Common *Rana temporaria*
Hedgehog *Erinaceus europaeus*
Lizard, Sand *Lacerta agilis*

Mouse, Wood *Apodemus sylvaticus*
Rabbit *Orycotolagus cuniculus*
Snake, Smooth *Coronella austriaca*
Squirrel, Grey *Sciurus carolinensis*
Toad, Common *Bufo bufo*
Vole, Bank *Clethrionomys glareolus*

INVERTEBRATES (Scientific names of Lepidoptera are given at first mention in the text)

Bug, Bed *Cimex lectularius*
Flea, Human *Pulex irritans*
Glow-worm *Lampyris noctiluca*
Hornet *Vespa crabro*

Ladybird, Eyed *Anatis ocellata*
Ladybird, Larch *Aphidecta obliterata*
Signal Crayfish *Astacus aquatilis*

References

Agassiz DJL (1996) Invasions of Lepidoptera into the British Isles. In Emmet AM (1996).

Agassiz DJL and Tuck KR (1999). The Cypress Tip Moth *Argyresthia cupressella* Walsingham 1890 (Lepidoptera: Yponomeutidae) New to Britain. *Entomologist's Gazette* 50: 11–16.

Albertini M, Damant C, Hall P and Halls J (1997). The Striped Lychnis moth, *Shargacucullia lychnitis* (Rambur, 1833) (Lepidoptera: Noctuidae) – a review of its distribution in Buckinghamshire during 1996. *Entomologist's Gazette* 48: 157–163.

Allan PBM (1943). *Talking of Moths*. The Montgomery Press, Newtown.

Allan PBM (1947). *A Moth Hunter's Gossip*. Watkins and Doncaster, London.

Allan PBM (1948). *Moths and Memories*. Watkins and Doncaster, London.

Allan PBM (1949). *Larval Foodplants*. Watkins and Doncaster, London.

Bailey M (1998). Good News for Spectacle Wearers. *Atropos* 5: 75.

Baker CR (1983). *Pest Species*. In Heath J (1983).

Barbour DA and Young MR (1993). Ecology and Conservation of the Kentish Glory moth (*Endromis versicolora* L.) in eastern Scotland. *The Entomologist* 112: 25–33.

Bourn N and Warren M (1997). The Road from Rio. *Butterfly Conservation News* 65: 7–9.

Bourn N, Green D and Parsons M (eds) (1999). *UK Biodiversity Action Plans: Priority Moth Species*. Butterfly Conservation, Colchester.

Bradley JD (1998). *Checklist of Lepidoptera recorded from the British Isles*. Bradley, Fordingbridge.

Bradley JD, Tremewan WG and Smith A (1973). *British Tortricoid Moths. Tortricidae: Tortricinae*. Ray Society, London.

Bradley JD, Tremewan WG and Smith A (1979). *British Tortricoid Moths. Tortricidae: Olethreutinae*. Ray Society, London.

Brooks M (1991). *A Complete Guide to British Moths*. Jonathan Cape, London.

Brown D (1997). County Focus – Lepidoptera in Warwickshire (V.C. 68). *Atropos* 3: 16–22.

Campbell B and Lack E (eds.) (1985). *A Dictionary of Birds*. Poyser, Calton.

Cooper MR and Johnson AW (1984). *Poisonous Plants in Britain*. HMSO, London.

Covell CV (1984). *A Field Guide to Eastern Moths*. Houghton Mifflin, Boston and New York.

Coyne JA (1998). Not black and white. *Nature* 396: 35–36.

Crafer T (1998). Ultra Violet Light and the Danger to Your Eyes. *Atropos* 4: 17–18.

Cramp S (ed.) (1985). *The Birds of the Western Palearctic.* Vol. IV. Oxford University Press.

Dandy JE (1969). *Watsonian Vice-Counties of Great Britain.* The Ray Society, London.

Davey P (1999). Weather conditions leading to the 1998 Green Darner *Anax junius* (Drury) influx. *Atropos* 6: 8–12.

Dickinson R, Asby R and Yates J (1999). What's new in Nature and Archaeology. *Sanctuary: Ministry of Defence Conservation Magazine* 28: 48.

Duckworth WD (1983). Introduction. In Watson A and Whalley PES (1983).

Dunbar D (ed.) (1993). *Saving Butterflies.* The British Butterfly Conservation Society, Colchester.

Emmet AM (1991). *The Scientific Names of the British Lepidoptera.* Harley, Colchester.

Emmet AM (1992). Chart showing the Life History and Habits of the British Lepidoptera. In Emmet A and Heath J (1992).

Emmet AM (ed.) (1996). *The Moths and Butterflies of Great Britain and Ireland,* vol. 3. Harley, Colchester.

Emmet AM and Heath J (eds.) (1992). *The Moths and Butterflies of Great Britain and Ireland,* vol. 7(2). Harley, Colchester.

Ennos RA and Easton EP (1997). The Genetic Biodiversity of Scottish Plants. In Fleming *et al.* (1997).

Fleming LV, Newton AC, Vickery JA and Usher MB (eds) (1997). *Biodiversity in Scotland: Status, Trends and Initiatives.* The Stationery Office, Edinburgh.

Ford EB (1945). *Butterflies.* New Naturalist. Collins, London.

Ford EB (1955). *Moths.* New Naturalist. Collins, London.

Ford TH, Shaw MR and Robertson DM (2000). Further host records of some West Palearctic Tachinidae (Diptera). *Entomologist's Record and Journal of Variation* 112: 25–36.

Fry R and Waring P (1996). *A Guide to Moth Traps and Their Use.* The Amateur Entomologist's Society, London.

Goater B (1986). A new technique of sugaring. *Entomologist's Record and Journal of Variation* 98: 37.

Goater B (1992). Drepanidae. In Emmet AM and Heath J (1992).

Goulson D and Entwhistle PF (1995). Control of diapause in the antler moth, *Cerapteryx graminis* (L.) (Lepidoptera: Noctuidae). *The Entomologist* 114: 53–56.

Grant BS, Cook DA, Clarke CA and Owen DF (1998). Geographic and Temporal Variation in the Incidence of Melanism in Peppered Moth Populations in America and Britain. *The American Genetic Association* 89: 465–471.

Hadley M (1984). *A National Review of British Macrolepidoptera. Invertebrate Site Register.* Unpublished Report 46. Nature Conservancy Council, London.

Haggett G (1998). Viper's Bugloss *Hadena irregularis* Hufn. and Spotted Sulphur *Emmelia trabealis* Scop. R.I.P. *Atropos* 5: 20–23.

Haggett GM and Smith C (1993). *Agrochola haemitidea* Duponchel (Lepidoptera: Noctuidae) new to Britain. *Entomologist's Gazette* 44: 183–203.

Heath J (ed.) (1983). *The Moths and Butterflies of Great Britain and Ireland,* vol. 1. Harley, Colchester.

Heath J and Emmet AM (eds) (1979). *The Moths and Butterflies of Great Britain and Ireland,* vol. 9. Curwen Books, London.

Heath J and Emmet AM (eds) (1983). *The Moths and Butterflies of Great Britain and*

Ireland, vol. 10. Harley, Colchester.

Heath J and Emmet AM (eds) (1985). *The Moths and Butterflies of Great Britain and Ireland*, vol. 2. Harley, Colchester.

Henwood BP (1997). Larval instars of the Lithosiinae (Lepidoptera: Arctiidae). *Entomologist's Gazette* 48: 257–258.

Hudson WH (1923). *Nature in Downland*. Dent, London.

Karsholt O and Razowski J (eds) (1996). *The Lepidoptera of Europe*. Apollo Books, Stenstrup.

Kettlewell B (1973). *The Evolution of Melanism*. Clarendon, Oxford.

Kuchlein JH and Ellis WN (1997). Climate-induced changes in the microlepidoptera fauna of the Netherlands and the implications for nature conservation. *Journal of Insect Conservation* 1: 73–80.

Leverton R (1998). *Eupithecia indigata* Hb. (Lep: Geometridae) larva eating aphids. *Entomologist's Record and Journal of Variation* 110: 80–81.

Leverton R, Young MR and Barbour DA (1997). *Epione paralellaria* D. and S. (Lep.: Geometridae) and its association with Aspen *Populus tremula* in the Scottish Highlands. *Entomologist's Record and Journal of Variation* 109: 49–55.

Maes D and van Sway CAM (1997). A new methodology for compiling national Red Lists applied to Butterflies (Lepidoptera, Rhopalocera) in Flanders (N. Belgium) and the Netherlands. *Journal of Insect Conservation* 1: 113–124.

Majerus MEN (1998). *Melanism: Evolution in Action*. Oxford University Press.

Manley WBL (1973). A guide to *Acleris cristana* (D. and S.) (Lep.: Tortricidae) in Britain. *Entomologist's Gazette* 24: 89–206, 4–7.

Marren P (1998). The English Names of Moths. *British Wildlife* 10: 29–38.

Morris MG (1983). Conservation and the Collector. In Heath J (1983).

Morris RKA and Collins GA (1991). On the Hibernation of Tissue Moths *Triphosia dubitata* (L.) and the Herald Moth *Scoliopteryx libratrix* (L.) in an old fort. *Entomologist's Record and Journal of Variation* 103: 313–321.

Musgrove AJ and Armitage M (2000). Rediscovery of the Narrow-bordered Bee Hawk-moth *Hemaris tityus* (L.) (Lep.: Sphingidae) in Breckland. *Entomologist's Record and Journal of Variation* 112: 75–76.

Oates M (1996). The Demise of Butterflies in the New Forest. *British Wildlife* 7: 205–216.

O'Connor RJ (1984). *The Growth and Development of Birds*. A Wiley – Interscience, Chichester.

O'Connor RJ and Shrubb M (1986). *Farming and Birds*. Cambridge University Press, Cambridge.

Pennington M (1996). *The Shetland Lepidoptera Report for 1995*. Shetland Entomological Group.

Pittaway AR (1993). *The Hawkmoths of the Western Palearctic*. Harley, Colchester.

Pollard E and Yates TJ (1993). *Monitoring Butterflies for Ecology and Conservation*. Chapman & Hall, London.

Porter J (1997). *Colour identification guide to Caterpillars of the British Isles*. Viking, Middlesex.

Pratt C (1981). *A History of the Butterflies and Moths of Sussex*. Booth Museum, Brighton.

Rosie JH (1976). Some Notes on the Macrolepidoptera of Caithness. *Entomologist's Gazette* 27: 13–26.

Rothschild M (1985). British Aposematic Lepidoptera. In Heath J and Emmet AM (1985).

Rutherford CI (1994). *Macro-moths in Cheshire 1961 to 1993*. The Lancashire and Cheshire Entomological Society.

Sailor RI (1983). History of Insect Introductions. In Wilson C and Graham C (1983).

Shaw, MR (1997). *Rearing Parasitic Hymenoptera*. The Amateur Entomologist's Society, London.

Shaw M and Askew RR (1983). *Parasites*. In Heath J (1983).

Shirt DB (1987). *British Red Data Books: 2 Insects*. Nature Conservancy Council, Peterborough.

Skinner B (1998). *Colour identification guide to Moths of the British Isles*. Viking, Middlesex.

Snow B and Snow D (1988). *Birds and Berries*. Poyser, Calton.

South R (1907–09). *The Moths of the British Isles*. Series 1 and 2. Frederick Warne, London.

Spence BR (1991). *The Moths and Butterflies of Spurn*. Spurn Bird Observatory.

Sterling PH and Langmaid JR (1998). The life history of *Acrolepiopsis marcidella*. *Entomologist's Gazette* 49: 151–154.

Tremewan WG (1985). Zygaenidae. In Heath J and Emmet AM. (1985).

Tutt JW (1901–05). *Practical Hints for Field Lepidopterists*. Elliot Stock, London.

Tutt JW (1902). *British Moths*. Routledge, London.

Waring P (1990). Essex Emerald Moth *Thetidia smaragdaria maritima* Prout (Lep.: Geometridae) – an Update. *Entomologist's Record and Journal of Variation* 102: 71–73.

Waring P (1993). Annotated List of the Macro-moths recorded in the British Isles, showing the Current Status of each species. *National Moth Conservation Project: News Bulletin 5*. Butterfly Conservation, Colchester.

Waring P (1995a). 'Wine-roping' for Moths. *Butterfly Conservation News* 60: 23.

Waring P (1995b). Wildlife Reports: Moths. *British Wildlife* 6: 393–395.

Waring P (1996a). Wildlife Reports: Moths. *British Wildlife* 8: 120–122.

Waring P (1996b). Wildlife Reports: Moths. *British Wildlife* 7: 324–326.

Waring P (1997). National Moth Conservation Project. *Butterfly Conservation News* 64: 16–17.

Waring P (1998). Wildlife Reports: Moths. *British Wildlife* 9: 324–326.

Waring P (1999). The National Recording Network for the Rarer British Moths. *Atropos* 6: 19–27.

Waring P (2000). Conserving the Barberry Carpet moth. *British Wildlife:* 11: 175–182.

Watson A and Whalley PES (1983). *The Dictionary of Butterflies and Moths*. Peerage Books, London.

Webb KF (1984). The Status of the Pine Hawk *Hyloicus pinastri* L. in Bedfordshire. *Entomologist's Record and Journal of Variation* 96: 53.

West BK (1993). *Orgyia antiqua* Ochs. (Lep.: Lymantriidae): Voltinism. *Entomologist's Record and Journal of Variation* 105: 241–242.

Wilson C and Graham C (eds) (1983). *Exotic Plant Pests and North American Agriculture*. Academic Press, New York.

Young MR (1997). *The Natural History of Moths*. Poyser, London.

Young MR and Smith R. (1997). The Rediscovery of *Ethmia pyrausta* (Pallas, 1771) (Lepidoptera: Ethmiidae) in Britain. *Entomologist's Gazette* 48: 85–87.

Subject Index

Species Index – *Lepidoptera*

(Species mentioned only in the appendices are excluded.)

Species Index – Other Groups